IISS

STRATEGIC
SURVEY
1997/98

Published by  Oxford
University
Press

The International Institute for Strategic Studies
23 Tavistock Street
London WC2E 7NQ

Strategic Survey 1997/98

Published by Oxford University Press for
The International Institute for Strategic Studies
23 Tavistock Street, London WC2E 7NQ

Director   Dr John Chipman
Editor   Sidney Bearman

This publication has been prepared by the Director of the Institute and his Staff, who accept full responsibility for its contents, which describe and analyse events up to 26 March 1997. These do not, and indeed cannot, represent a consensus of views among the world-wide membership of the Institute as a whole.

Assistant Editor: Maps   Dr Mats R. Berdal
Managing Editor   Susan Bevan
Research Assistant   Ellen Peacock

Design and Production   Mark Taylor

First published April 1998

ISBN   0 19 829420 4
ISSN   0459-7230

Strategic Survey (ISSN 0459-7230) is published annually by Oxford University Press.

The 1998 annual subscription rate is: UK £27.00; overseas $US44.00.

Payment is required with all orders and subscriptions. Prices include air-speeded delivery to Australia, Canada, India, Japan, New Zealand and the USA. Delivery elsewhere is by surface mail. Air-mail rates are available on request. Payment may be made by cheque or Eurocheque (payable to Oxford University Press), National Girobank (account 500 1056), credit card (Access, Mastercard, Visa, American Express, Diner's Club), direct debit (please send for details) or UNESCO coupons. Bankers: Barclays Bank plc. PO Box 333, Oxford, UK, code 20-65-18, account 00715654.

Claims for non-receipt must be made within four months of dispatch/order (whichever is later).

Please send subscription orders to the Journals Subscription Department, Oxford University Press, Great Clarendon Street, Oxford, OX2 6DP, UK. *Tel* +44 (0) 1865 267907. *Fax* +44 (0) 1865 267485. *e-mail* jnlorders@oup.co.uk

Strategic Survey is distributed by M.A.I.L. America, 2323 Randolph Avenue, Avenel, New Jersey, NJ 07001, USA. Periodical postage paid at Newark, New Jersey, USA and additional entry points.

US POSTMASTER: Send address corrections to *Strategic Survey*, c/o M.A.I.L. America, 2323 Randolph Avenue, Avenel, New Jersey, NJ 07001, USA.

PRINTED IN THE UK by Bell & Bain Ltd, Glasgow.

Contents

List of Tables and Figures

Perspectives

Two very different crises blew up in 1997 and early 1998, on different sides of the world and of very different dimensions. One was tiresomely familiar: a political stand-off between the mendacious regime in Iraq and the UN. The other was as unexpected as it was dangerous: a financial and economic collapse of at least four of the hitherto highly successful Asian Tigers. Different as these crises were, they confirmed once more that it is the attitude and power of the United States which shape the world's reactions to major crises. The combination of its military power and its economic and financial clout gives the US unique weight. Only the US has the capacity to lead, and, when it wishes to exercise the capacity to the full, it is able to dictate the terms on which solutions can be found.

In the current political and economic climate the US has gone beyond being the world's only super-power to become, as French Foreign Minister Hubert Vedrine has said, a 'hyper-power'. Yet its power is constrained at times by the exigencies of US domestic politics, and at other times by the tensions arising from the conflict between its innate idealism and the often necessary exercise of *Realpolitik*. Furthermore, while US power can influence the role of international institutions, the need to work through them can sometimes alter the terms on which crises are solved.

For the management of serious crises in the world, it is always necessary for the US to act. Unfortunately, its reactions are sometimes late because it is slow to recognise the strength of a developing crisis and slow also to decide how heavily US national interests are engaged. This difficulty has been compounded by globalisation, which has created conditions within which crises breed with increasing speed, but also, initially with uncertain impact. The sources of these problems are so various that even the US, fully employing all its resources, can not be certain to solve them. Unless the US chooses to frame a policy, however, it is unlikely that any answers will be found to the security crises thrown up by globalisation. This was all the clearer in a year in which most of the major European allies of the US were to a large degree concentrating on their own economies in order to prepare for entry into a single currency, the euro.

These trends point up some enduring truths. While crises develop, as they are doing with greater frequency, the Clinton administration's attention is always being diverted by its deep concern about the possible reaction of a hostile Congress to any development. Even at the best of times, US attention to the world outside is intermittent. This in itself contributes to the nurturing of crises. Since early or persistent action is rare because of the administration's limited power to concentrate on foreign policy, problems can be allowed to brew excessively. And only when they reach crisis proportions is it possible to argue, to a still relatively indifferent public, that US power needs to be deployed forcefully.

The financial difficulties in Thailand were widely understood before the run on the Thai *baht*, but the US did not prod international financial bodies, like the International Monetary Fund (IMF) and World Bank, to act until the collapse of Thailand's economy had dragged down some of the other East Asian Tigers. Saddam Hussein's obstruction of the international inspection teams had increased steadily throughout 1997, but it was not until the crisis reached boiling-point that the US moved decisively to force him to back down.

In both cases, the US decision to work through multinational institutions to achieve its ends meant that the action taken was not as clear-cut as it might have been. It is difficult for these bodies to act single-mindedly in a crisis when so many states can influence policy. The UN, although helpful in averting the use of military force against Iraq, diluted the clear stance championed by the US and the UK. This has opened up the possibility that Iraq will wriggle out from under the sanctions regime before its strongest critics are convinced it has met the conditions established for lifting the sanctions. While the international financial institutions ended 1997 with pledges of well over $100 billion to put the Thai, South Korean and Indonesian economies back on the road to growth, the whole package was arrived at in hesitant stages and many critics felt this may have slowed the prospect of recovery.

The Suffering Tigers

The collapse of financial markets in a number of East Asian countries also collapsed at least one shibboleth of our time. This was the view, vigorously put forward by a number of Asian leaders, that an ineffable 'Asian Way', distinct from Western capitalist methods and not reproducible in the West, had helped ensure the earlier success of their economies. Unfortunately, it was the very essence of this Way which helped create the economic crisis of 1997. Indiscriminate bank lending on poor, or even no, collateral, tight government control of the markets, crony capitalism, a refusal to allow banks or businesses to go bankrupt, and an insistence that everyone was entitled to hold their job for life all contributed to the conditions which created the crisis.

As part of the rescue packages organised under IMF auspices the suffering countries are being pushed to change their approach in each of these areas. From the point of view of the Western nations who supplied the rescue funds it seems reasonable to proclaim 'no change, no funds'. They are probably right to do so. If these states wish to recover the position they held before the collapse they will need to reinvigorate international, as well as domestic, confidence in how their economies are run. To them, however, this looks very much like rich Western nations taking advantage of their present temporary embarrassment to enforce alien methods of operation to the benefit of Western businesses and markets.

South Korea and Thailand are the furthest advanced in instituting the necessary changes (although as yet with neither the thoroughness nor speed that will ensure a quick recovery). The IMF money is flowing in and the mid-term future looks better than it did six months ago. Indonesia, however, has been acting like a bride unsure if she wishes to appear at the altar. President Suharto, the 74 year old architect and (with his family) the greatest beneficiary of the once-booming economy, was tenacious in his efforts to avoid following IMF advice on the need for economic restructuring to avoid a complete collapse into depression.

US pressure, applied through a series of high-level visits to Jakarta by Treasury and Trade officials and several long telephone calls by President Clinton, seems to have headed off what would have been a fruitless attempt to support the Indonesian *rupiah* through a so-called currency board. IMF threats not to deposit the $3bn tranche of aid, due towards the end of March 1998, undoubtedly also helped. Yet the situation remains delicate. There is no assurance that either Suharto, 'elected' in mid-March 1998 to his seventh five-year term as President, nor his probable successor Vice-President B.J. Habibie, are willing to follow the prescription of the West's economic doctors. Indonesia will have to be carefully nursed along under constant watchful care to ensure that a further deterioration in its economy does not affect the other weakened economies in Asia.

Globalisation and Security

It was the fear that neighbouring economies could not but be affected by the run on devalued currencies and the attendant economic collapses that led the West to rush in with its money and advice. Well before the rescue could reasonably be thought effective, however, it was clear that there would not be an extended domino effect. It was a mistake to believe that 'globalisation', in its very much over-hyped form, meant there was such a firm interconnection and interdependence of markets and money flows that what happened in one area would perforce have a similar affect on others. The US and European markets have not been adversely affected by the crisis; rather they have risen to unprecedented heights. Even economies closer to the whirlpool have not been sucked in. China has acted carefully in not devaluing its currency, which would have hit the regional economies with another blow, and has itself been insulated from the worst effects of the crisis. The Japanese economy has slipped into the doldrums, but this is a long-standing condition and its markets have not plummetted in reaction to the current crisis.

Globalisation is an uneven work still in progress. Where it has taken hold, in the technologically advanced and economically developed parts of the world, it is a potent, but not overwhelming force. There, it is basically an economic and communications phenomenon. It is the search by investment capital for the most cost-effective, productive and profitable investment opportunities, regardless of their location and the partners involved. For money managers, it is the search for the quickest, highest return on currency values through global trading. For all concerned, governments, business-men and even the general public, it is the use made of the increasingly cheap computer, information and communications technology to reach around the globe and into each others' societies almost at will.

This is not a wholly new phenomenon. There have been historic eras, particularly in the days of the British empire, when trade was less constrained than it is today and investment capital flowed freely and globally. The invention of steamships, the telegraph and the telephone all suggested to commentators of those days that the world had become smaller. What is altogether new, however, is the speed with which funds and information now move globally, the depth to which they penetrate societies and the extent to which these developments have eroded the capacity of governments to govern and to control what enters their borders from outside (or what passes out through them).

Ironically, just as a large number of new states has achieved ostensible national sovereignty, their command of the tools of economic authority has disappeared – not into the hands of larger and more powerful mercantilist states, but into a black hole where thousands of private sector investment and financial decisions exert a force which is apparently unaccountable. The daily turnover of leading currency markets now far exceeds the global stock of foreign exchange reserves. The powerlessness of governments, in the face of sudden large movements of capital, rapid decline in value of local business assets and stocks and the collapse of many of these businesses, suggests that the open winds of the global economy can blow bad, as well as good, through a nation's economy and political stability.

Economics is only one dimension through which globalisation can have an impact on international security. The widening capacity to make sophisticated weapons is clearly another important dimension. The arms making industry has begun to go global, with transnational cooperation in arms manufacture and increasing integration of new warfare technology with global commercial technology. The proliferation of technological know-how and the capacity to manufacture modern weapons significantly increases the threat to security worldwide. A rapidly increasing number of states are also achieving the technical level to undertake research into nuclear, chemical and biological weapons and they can acquire the necessary parts through commercial channels.

The global economy exacerbates other security threats. Terrorists have increasing access to communications and information technologies which allows them to organise on a worldwide basis. International crime and narcotics syndicates have 'gone global' with the ability to move money around the world almost instantaneously and to communicate easily over long distances, thus evading the capacity of individual governments to detect and control the threat. All these developments have raised fundamental questions about the capacity of the international system, as currently organised, to ensure stability and security. The challenge to the managers of modern security is to find, either through new institutions or new methodology, a way to create and co-ordinate sound global policies to control the emerging threats.

Old Threats and Old Machinery

Even the machinery for dealing with old and familiar threats has begun to creak. Saddam Hussein has spent the seven years since the Gulf War in 1991 evading the efforts of the UN Special Commission (UNSCOM) to probe his weapons of mass destruction programme. UNSCOM managed, despite these efforts, to dismantle his nuclear weapons programme and has taken measures to place high barriers in the way of any attempt to re-create such a programme. Iraq's biological and chemical weapons, however, are a very different matter. UNSCOM's experts have uncovered a vastly more sophisticated and extensive research and production programme for these weapons than had been expected. They are convinced that there is a great deal more to learn. But unity on the Security Council is an essential support to UNSCOM's unwelcome attention to Iraq's perfidy, and it is wasting away.

Saddam Hussein seemed to have felt that the atrophy had gone so far that a sharp push would tip the balance within the Security Council to his advantage. Both France and Russia, mainly for commercial reasons, and China, for ideological ones, argued for a slackening of the sanctions regime. The official position of the US was that sanctions must remain at least until UNSCOM was satisfied. Unofficially it was clear that the US felt the pressure must be kept high until Saddam Hussein was deposed. This discordant tune provided the background for Saddam's latest dance of the seven veils.

In one sense he miscalculated once again. He thought the lack of an international consensus would force the US to accept a compromise in his favour. In the first round he seemed to have guessed correctly. When the US threatened action in November 1997, because Saddam had refused to allow inspectors, and particularly those from the US, to carry out their allotted work, Russia's Foreign Minister Genady Primakov rushed to Baghdad and brokered a loose agreement that was a definite step back, but which was reluctantly accepted by all as a useful diplomatic solution. This seems merely to have spurred Saddam to pick up the tempo. He reneged on the agreement, as he had with so many others, and openly confronted the United States, now fully aroused.

Here again the US was late to see the depths of the crisis, but once recognising it, moved to decisive action – action no other nation could have taken. Although the US was frustrated in the early days of the stand-off by its inability to gather around itself once again the coalition that President Bush had commandeered in 1990/1991, it made clear that, if necessary, it would take military action alone. In the event, the UK was prepared to join the fray, sending an aircraft carrier to the Gulf and pledging its full support. The force the US mustered was massive; US threats of action escalated. Although this was not the preferred course for the Clinton administration, it seemed driven to use heavy military force in pursuit of its goals.

The threats galvanised diplomatic activity. A steady stream of foreign ministers flowed to Baghdad. France and Russia were joined by a number of Arab leaders in the effort to avert the threatened massive destruction of elements of Saddam's power base. For Saddam it was a rare moment; courted by world leaders, he strutted before the television cameras, milking the situation for all its propaganda value. At the eleventh hour, he even had the Secretary-General of the United Nations, Kofi Annan, negotiating with him. Although Annan had come bearing the demands of the US and the UK, and had succeeded in getting Deputy Prime Minister Tariq Aziz to sign an undertaking to open all suspect sites to UNSCOM (subject to the provision of greater diplomatic cover for the inspection of presidential sites), the politics of the situation had shifted. The pressure was now on the West to find a more effective policy to keep Iraq's ability to develop weapons of mass destruction under control. The sanctions regime had become a weak reed, and more a liability than a useful tool.

By keeping Iraq from reacquiring the means of threatening its neighbours the US has been ensuring the security of those nations. Yet during this crisis all but Kuwait distanced themselves from the US. For most regional states the reality of the Iraqi threat had dissipated. Moreover, they had seen that military action, which had been used before in the years since the end of the Gulf War, had done nothing to weaken Saddam's grip on power, nor to bring him into greater compliance with UN strictures. In their view sanctions had not done so either. Both merely punished the poor and defenceless in Iraq and the US-led policy was destabilising the domestic politics of much of the region. Anti-Americanism had been aroused because of the stalled Middle East peace talks. No Arab regime was keen to support military force that would have an uncertain effect, only to pay the price of more domestic dissent, especially from Islamic radicals. The whole regional diplomatic game suggested that the long-standing desire to achieve regional solutions to regional problems had reasserted itself. During this period, the Iraqi foreign minister was received by Syria for the first visit at this level for 17 years. Saudi Arabia and Iran began to resolve some of their differences. The US is in danger in this context of losing vital regional support for its general Gulf policy.

In any case, the tensions that have developed between the intrusive inspection regime and the sanctions regime can not be maintained. No one imagined that both sanctions and inspections would last seven years. In the short-term, with the agreement brought back from Baghdad by Kofi Annan, inspections will be allowed so long as they do not threaten to uncover what Saddam wishes to hide. In the mid-term, one or the other will probably have to go. Although it will be difficult to find a way to insist on intrusive inspections without the cudgel of sanctions, it will be necessary to do so. Before that is possible, it will be vital for the US once again to link its Gulf policy with the peace process, as it had so successfully in 1991.

Time To Lean Hard

For over a year, relations between Israel and the Palestinians have been frozen. The atmosphere has become more poisoned than at any time since the Oslo Accords in 1993. The Likud government's refusal to carry out the withdrawals according to the schedule reached in the Oslo II agreement reflects the weakness of Netanyahu's grasp on power. Small religious and far-right parties which hold the balance of power insist that Israel retain its hold on all the land it now occupies on the West Bank, and either build or expand settlements which narrow the areas the Palestinians might someday acquire. Whether Netanyahu agrees wholeheartedly or not is almost beside the point. He wishes to survive as prime minister. To do so he must placate them.

In the Arab view, the loss of momentum in the peace process is wholly the responsibility of the Likud government, with its intransigent position. While it is true that Yasser Arafat has not completely carried out his part of the agreement (particularly with regard to changing the PLO charter, so that it no longer calls for the destruction of Israel, or reducing virulent anti-Israeli propaganda), the more extreme demands that Israel makes are beyond his power. He could not, if he wanted, completely control terrorist acts. He could not wipe out the *Hamas* movement without destroying himself as well. He has shown himself willing to move a considerable distance towards accommodating Israeli demands because, if there is no movement towards peace, his political life will undoubtedly end ignominiously.

In these circumstances, only a change in the position of the strongest party to the negotiations can restart the talks. Unless Israel can be brought back to the promising path it was following under the previous government, there can be little hope. The real danger is that the current uneasy equilibrium cannot hold. If the talks are not restored in an atmosphere of at least a minimum of trust, increasingly serious violence is almost certain to break out. Whether this would escalate to a full scale *intifada*, perhaps even one led by Arafat and the PLO, is less certain. But the risk is so high that other avenues must be explored quickly to avert it.

Most observers, and certainly the Arabs, would assert that only the US has the influence and power to put the peace train back on the track and get it moving again. That it does not seem willing to use that influence to the full has dissipated its moral leadership in the Arab world. The US has engaged in the usual rounds of discussions with the Israelis, first at the level of Special Envoy Dennis Ross, through Secretary of State Madeleine Albright to the President. At one point, in November 1997, US anger at positions taken by Israel led to a presidential snub while Prime Minister Netanyahu was in the US. That did not seem to affect Netanyahu's position in Israel adversely. In February 1998, individual presidential meetings with Arafat and Netanyahu did nothing to advance the process. Israeli concern that the US is prepared to make public its own formula for breaking the stalemate – a call for a withdrawal of Israeli forces from at least 13% of the West Bank – led it to take an unusual preventive action by announcing that the cabinet had unanimously rejected the rumoured formula even before it had been advanced.

If the US allows itself to be put off by this action, it will have lost much of its remaining credibility and influence in the region. That credibility is already being dissipated. This is as true with the Israelis as with the Arabs. Patience in diplomacy, as in most aspects of life, is a virtue; but carried to extremes can turn into a vice. There comes a time to act, and, to avert the consequences that would follow a complete collapse of the talks, now is the time.

The Self-Hobbled Giant

There is no simple explanation for the conundrum that the US is the only super-power in the world, with overwhelming economic and military power, and yet finds it so difficult to use that power to achieve its goals. One reason lies in the structure of the US government. The founding fathers, engaged in a battle to rid themselves of an autocratic ruler, carefully wrote a set of checks and balances into the constitution which would prevent any President from becoming a King. They did their work well, perhaps even too well. Paradoxically, the presidency, often called the world's most powerful position, is in fact a weak one. UK Prime Minister Tony Blair, in command of a huge majority in Parliament, can be sure that the wishes of his cabinet will be carried through. But, as President Harry Truman said to his aides in the Oval Office just before General Dwight D. Eisenhower was inaugurated as president: 'He'll sit right here and he'll say do This, do That! And nothing will happen. Poor Ike – it won't be a bit like the Army.'

The President must always be concerned about what Congress will think of any projected action. This is doubly true when control of Congress is not in the hands of his own party. Congress can block the President's choices of ambassadors or special representatives, can prevent treaties he has negotiated and signed from coming into being, and can provide bills of its own which force him to take actions he is convinced are contrary to the interests of the nation. The President must always be concerned not only about what effect an action will have on foreign countries, but also on the effect it will have domestically. He is dependant on the mood of the people, and the mood of their representatives on Capitol Hill.

It is sometimes said that US foreign policy is not consistent. The remarkable thing is how consistent it has been on many issues for many years. The greater danger is that it has often been too consistent; that, partly as a consequence of the need to placate Congress, it has been too inflexible to meet the changing needs of changing times. Strong lobbies have the ear of Congress, and the President must thus listen as well.

US support of Israel has been a constant for fifty years, almost irrespective of Israeli actions, or lack of action, that various US administrations opposed. Since the collapse of the Soviet Union, the desirability of a change in the US position on Cuba has been recognised by the administrations of both George Bush and Clinton. The best either could do was to put in motion very small and exceedingly gradual adjustments. The same is true with regard to Iran, despite a desire to adjust policy to encourage the changes taking place in that country.

This inflexibility is one of the major impediments to the effectiveness of US policies, not the putative lack of consistency. The problem has been compounded by the geo-strategic changes of the past decade. So long as there was a single dangerous enemy in the form of the Soviet Union, lack of strategic flexibility was not a serious problem. It was enough to have a single over-arching strategic aim and to act within it with tactical flexibility. When that more dangerous, but strangely more comfortable, era ended, it was no longer enough to rely on tactical adroitness alone. The various pieces of the strategy no longer meshed easily and the difficulty of adjusting them, even with deliberate speed, was telling.

Even when the President can find a policy which will gain the support of Congress and thus allow the US to act unilaterally, there are prices to pay outside the US. Inevitably, action or even inaction by the powerful will breed resentment among the weak. US actions, sometimes taken much more for the benefit of the recipient as in the case of the Arab neighbours of Saddam Hussein, can often induce an anti-American backlash. Many Asians, including some leaders of the nations involved, condemned the West and particularly the Americans even as they were providing the means by which the Asians could surmount their own crisis.

The Hope of Multinational Institutions

Either because it is hamstrung by Congress, or because it prefers to avoid a backlash, the Clinton administration has leaned in the direction of using international institutions as the instrument of its foreign policies. This is encouraged by other states, such as France and Russia which seethe with frustration because they no longer have the power to act unilaterally. Having encouraged their voice to be heard, however, the US must soften its own, and its policies are frequently diluted as a result. This arouses those in Congress whose vision is restricted to a single issue, and US support for the international body is compromised. The inability of Congress to appropriate the more than $1bn needed to pay the US arrears to the UN is due to the extremism in the House, and the successful battle one Representative waged on the abortion issue. Unless the President is prepared to sign a law which would require any private international recipient of US aid funds to abjure support for abortion no money will be approved by Congress for the UN. Result: weak institutions are left even weaker.

Nor can the US rely on willing coalitions to provide support for action. Allies can be bought off. The siren song of commercial gain is a strong competitor to unity. Every international action of significance taken by the US requires specially tailored negotiations with key allies if these are to carry the greatest possible weight and thus strip US initiatives of their unilateral character. Such negotiations are easiest with international bodies such as NATO where the US leads unambiguously, but are harder elsewhere. The US is bound to find itself often in the future balancing the benefits of a more multinational approach (strength of international resolution, assistance in the management of crises) against the utility of a unilateral approach which allows the US its preferred policy without the encumbrances of inter-allied consultation. The quality of US leadership in the future is likely to be judged by the wisdom of the choice it makes between these mutually exclusive methods for dealing with crises.

Strategic Policy Issues

Year Of The Sick Tigers

Like the collapse of Soviet-backed regimes in Eastern Europe and the break-up of the Soviet Union itself, the unravelling of the 'East Asian Economic Miracle' astonished the world with its suddenness. What started in early 1996 as a worry about a property bubble in Bangkok and a related malaise in the Thai banking system had, within a year, engulfed much of South-east Asia, as well as South Korea, in a vicious cycle of sharp currency depreciation, threatened national insolvency, severe fiscal austerity and, at best, a grinding economic slowdown, or, at worst, acute recession. Yet these were countries which had shown sustained rates of economic growth unprecedented in world history. They were widely believed to be immune from the boom-bust cycles that have afflicted other developing countries in, for example, Latin America.

Economic self-confidence had given rise to political assertiveness as well, and the belief that there was something special about Asian modes of social and economic organisation and of intra-regional diplomacy. Economic disaster will thus also have far-reaching political, ideological and strategic implications. They are certain to include domestic political change, increased intra-regional tension, a humiliating recognition of dependency on the United States in the short-term, and, in the long-term, the possibility of a nationalist backlash and a greater role for regional powers like China and Japan.

What Went Wrong?

In early 1997, there was still much dumbfounded admiration for the miracle, which appeared to have produced an unprecedented combination of sustained and rapid economic growth, increased social equity, and political stability. By then it was clear to many – although not apparently the Thai government – that Thailand was facing serious financial difficulties. Hardly anybody, however, foresaw the extent of that country's economic disaster, let alone its contagious effect elsewhere in the region. Thailand was the butterfly that, as chaos theory would have it, fluttered its wings and caused an earthquake .

The catalyst for the implosion was a regional downturn in exports, which particularly affected Thailand. After years of growing at close to

20% annually, exports from the region in 1996 showed no growth at all. Three explanations are offered for this phenomenon. Firstly, there was a cyclical fall in demand for electronic products. Secondly, there was mounting global over-capacity in many other industries as well, attributable in large measure to the increased economic activity in China. There, a sharp devaluation of the *renminbi* in 1994 had been followed by several years of remarkable success in bringing down the inflation rate. These two factors made Chinese exports extremely competitive.

Third and most important, however, was the breakdown of an international currency consensus in place since the 1985 New York Plaza Accord. That had allowed the yen to strengthen against the dollar, making Japan's exports more expensive, and encouraging its manufacturers to relocate capacity to continental Asia, where, since local currencies were linked to the dollar, production costs were cheap. The extent of this move was truly staggering. For example, in 1987, Thailand received more Japanese investment than in the entire course of the previous twenty years. Put simply: Japanese investment poured into the region and cheap exports poured out. But this happy arrangement began to unravel as the dollar appreciated – rapidly and sharply – against the yen. From its peak of about 80 yen to the dollar in 1995, the yen fell to a low of 129 to the dollar at one point in late 1997. Correspondingly, East Asian currencies, which were all linked in one way or another to the dollar, appreciated against the yen and most other currencies in the world. Initially, the wave of devaluations which then took place in these currencies merely restored their level against the yen.

The strains this shift produced reached breaking-point first in Thailand for a variety of reasons. Thailand was facing a structural as well as cyclical difficulty in its export industries. With a per capita gross domestic product (GDP) of around $2,000, it was pricing itself out of the market in labour-intensive industries. But it had yet significantly to alter its export composition into the higher value-added industries like electronics. Insolvency in the banking and finance company sector loomed closer in Thailand than in other countries. By early 1998, 56 out of 91 Thai finance companies had been closed down, and four out of 15 banks had been nationalised after finding themselves unable to repay liquidity support provided by the central bank.

In addition, Thailand operated the most rigid of all the exchange rate mechanisms in the region, with the exception of Hong Kong's currency board system (which is fundamentally different in that the Hong Kong dollar is fully backed by holdings of dollars). This firstly had the effect of keeping *baht* interest rates much higher than those paid for dollars, to protect the currency. Banks and companies could then borrow cheaply in dollars to fund *baht* assets at what appeared to be negligible real interest costs. Secondly, the rigidity of the band gave the central bank very little leeway in defending the currency. Interest rates could be pushed sky-high, billions could be spent buying dollars, and capital controls could be imposed. At various times, all three methods were tried, sometimes simultaneously. But, after the Bank of Thailand had nearly bankrupted itself in the forward market, it was forced to give way and allow the *baht* to float. Instead it sank. Thailand became the first of four countries to turn to the International Monetary Fund (IMF), which arranged more than $100 billion in rescue funds for Thailand, Indonesia and South Korea, and smaller emergency credits for the Philippines (which had been operating under an IMF programme for more than thirty years).

In retrospect, it does not seem so surprising that other currencies should have toppled like dominoes in the storm that followed the *baht's* effective devaluation. Using the now fashionable economic measures of indebtedness, current account deficits and non-performing loans, the other countries in the 'ASEAN 4' (Thailand, the Philippines, Indonesia and Malaysia) also seemed vulnerable. Indonesia and South Korea shared Thailand's vulnerability to a dangerous volume of short-term foreign debt.

International currency speculators had spotted that there was money to be made in selling the *baht* short. More important, local companies which, like those in Thailand, had borrowed offshore and not hedged their exposure, began rushing to do just that, by buying dollars before their local currency depreciated. There was also a certain pressure for competitive devaluation. In thin and nervous markets, such perceptions were enough to bring the other regional currencies down too.

Change At Home

The effect was catastrophic. Currency falls of between 35% and 50% in Thailand, Malaysia, the Philippines and South Korea, and, at times, of 80% in Indonesia, pushed up the cost of foreign currency debts far beyond the ability of local banks and companies to service them. By February 1998, almost every company listed on the Jakarta Stock Exchange was, on paper, bankrupt. By then, what had begun as a financial crisis had become an economic crisis. Across the region, growth rates were falling fast. Thailand and Indonesia were in the grip of an alarming economic contraction. Factories were closing, construction projects were being abandoned, and unemployment and prices were rising.

The political consequences of this downturn were exacerbated by two factors: many governments had based their claim to legitimacy on the ability to deliver very rapid rates of economic growth; and in many countries a process of political change was, regardless of the financial turmoil, on the constitutional calendar. In some countries, these factors combined to entrench a fragile democratic process. For example, a new constitution was adopted by the Thai parliament in August 1997. Many of its provisions, aimed at reducing the role of money in the electoral process, were abhorrent to politicians. The economic climate, however, made them unable to oppose a set of popular reforms. In South Korea in December, Kim Dai Jung became the first opposition candidate to win a presidential election since the end of military rule in 1987. And in the Philippines, those manoeuvring for a constitutional amendment to allow President Fidel Ramos to stand for a second term in the elections due in May 1998 were forced to abandon the idea. The economic downturn militated against embarking on a divisive political campaign which opponents alleged would undermine the democratic gains of the 1986 people power revolution.

In Malaysia, too, the economy became politicised in the rivalry between supporters of the Prime Minister, Mahathir Mohamad and of his deputy, Anwar Ibrahim, who is also the Minister of Finance. Anwar is Mahathir's designated successor; their different approaches to handling the crisis – nationalistic and accusatory in Dr Mahathir's case, conventional and soothing in that of his deputy – fuelled the debate about when power should be transferred.

But it was in Indonesia that the political consequences of economic collapse were most drastic. This was partly because that economy suffered more than any other. But also, its political system was the most rigid, and the most obviously bound up with the causes of economic ruin, because of the commercial dominance of the family and friends of the president. After 32 years in power, President Suharto appeared determined to soldier on with a seventh five-year term, confirmed in largely ceremonial elections held by an assembly that convened in March 1998. Food shortages, spiralling prices and a history of antagonism towards the disproportionately wealthy Chinese minority contributed to widespread rioting. There were fears that the situation could turn even more violent, unleashing a flood of refugees, and possibly destroying Indonesia as a unitary nation. This naturally caused near panic among officials in neighbouring Singapore and Malaysia, who also feared that a replacement regime might be rather different from the inward-looking, secular Suharto era.

Regional Weakness

One of the reasons for the founding of the Association of South-east Asian Nations (ASEAN) in 1967 was to engage Indonesia in a peaceful regional club, and so bury for ever the memory of the period of confrontation between it and the Federation of Malaysia in the early 1960s. These origins help explain the importance ASEAN members attach to the principle of non-interference in each others' internal affairs. This mutual tolerance had served the organisation well in avoiding conflict, but by 1997 had proved inadequate to deal with a series of challenges from a coup in Cambodia to the appalling smog that smothered much of South-east Asia later in the year.

Most disappointing for ASEAN members, the organisation was unable to provide any effective form of mutual self-help in the face of the economic crisis. Dr Mahathir, in particular, pushed the idea of an Asian or ASEAN fund to provide balance-of-payments support for countries whose currencies were under attack. Japan's withdrawal of backing for the scheme, however, left it an empty shell, and it was replaced by a 'Manila framework' agreed in October 1997. This, in essence, left regional economic health-care in the hands of the IMF. The only substantive ASEAN initiative was a proposed regional surveillance mechanism that would help co-ordinate macro-economic policy and promote peer pressure. Such a mechanism would be useful. But it came too late. ASEAN had failed to persuade Thailand to tackle the economic troubles which sparked the crisis. It was unable to dissuade Mahathir in August and September from making speeches blaming foreign speculators and threatening retaliation, and hence aggravating the markets' antipathy to the region. And by February 1998, regional leaders were watching helplessly as President Suharto accepted the advice of an American economist, introduced by Suharto's children, who was prepared to peg the Indonesian rupiah by means of a currency board system, like that in Hong Kong. Other ASEAN members thought this could be disastrous for Indonesia, and hence, by extension, for themselves.

Mahathir was meanwhile promoting a more modest form of regional cooperation: the use of the local currencies in intra-regional trade. But many of his ASEAN colleagues believed in private this would be impracticable and, if not, would have only a marginal impact on exchange rates.

More Squabbles To Come

Although there was a frenzy of high-level meetings among ASEAN leaders and they maintained their unity, the limitations within that unity became clearer. Ironically, ASEAN has achieved a greater degree of economic cohesion in distress than in prosperity – the markets, at least, have treated the region as a whole, so economic policy in one has a direct effect on the others. As the region sets about recovery, this is likely to lead to further tension.

Another reason for believing there may be more rather than less friction over economic policy is the likelihood of increased competition for export markets. The best quick fix to the region's troubles relies on tapping the still healthy growth of American demand. This goes against the trend that has developed in recent years towards a rapid growth of intra-regional trade. With Japan and South-east Asia both in a period of slower growth, the pressure to increase market share in the US will mount.

The pressure will become particularly acute if the contagion spreads much further in North-east Asia. By early 1998, Taiwan as well as South Korea had seen its currency depreciate. The fear was that China, despite frequent denials, could be forced by a slowing domestic economy, mounting unemployment and, perhaps, social unrest to cheapen its currency, leading to another round of devaluation.

Migrant labour may well create a further aggravation. Malaysia has some 1.8 million foreign workers, mainly from Indonesia. Thailand has up to a million, mostly from Myanmar. Both host countries responded to the economic slowdown by threatening to deport large numbers of illegal immigrants. In late March, Malaysia began to force some recent illegal immigrants back to Indonesia. The threats to export large numbers, however, were probably designed more to placate domestic opinion than for imminent realisation. Yet a prolonged downturn will inevitably create shortages of jobs and potential trouble as foreigners are asked to return to homelands also in serious economic difficulty.

Friends In Need

The failure of the region to help itself, threw the role of saviour onto the IMF, and, in particular, its largest shareholder, the US. Both, inevitably, found themselves unpopular in some regional circles. The IMF was criticised for imposing a 'one size fits all' package of austerity measures, structural reforms and monetary policies regardless of the huge differences between the Asian economies and earlier beneficiaries of IMF help such as Latin America. The IMF was accused of ignoring the fact that these were governments with a record of fiscal prudence, and that the external payments crises were largely caused by private sector borrowing. Hence, the argument went, the budget surpluses and high interest rates prescribed by the IMF were unnecessary and likely to deepen the recession. When massive IMF intervention failed to halt the slide in late 1997, criticism of the organisation gathered ferocity. IMF officials argued that its critics were missing the point – the key element in its packages was not the fiscal targets (which in February 1998 were eased for Thailand), but the structural reform it was suggesting. It also pointed out that high interest rates were likely to be essential until currencies had stabilised. They were dictated by the market.

Meanwhile, the IMF was under fire in the US from liberals (appalled at its rescue of dictators like Suharto), from free marketeers arguing it was bailing out the Western banks which had financed the bubble economies, and from a domestic trade lobby afraid the Fund was going to cost American jobs by supporting export-led growth in its client countries, some of which were now super-competitive because of devaluation.

The US likewise found itself facing contradictory charges in Asia: that the crisis was a result of a conspiracy it had thought up to cut Asia down to size and that it was not doing enough to help. Mahathir articulated the conspiracy theory most angrily, repeatedly denouncing the hedge fund operator George Soros, who was seen as having close White House links. To most in the West, the idea of a concerted conspiracy seems preposterous, but it would be wrong to under-estimate the regional constituency for such arguments.

The feeling of being let down by the US was most acute in Thailand. A number of politicians pointed to Thai–US cooperation during the wars in Vietnam and the Gulf, and felt slighted now that their country was no longer of such strategic importance. The US insisted this was not so, and that it had played an active role in formulating the Thai package, and indeed had contributed directly to the Indonesian and South Korean rescues (seen in Thailand as confirmation that the militarisation of the Korean peninsula, the importance of Indonesian sea-lanes, and the greater size of, in particular, the South Korean economy, gave those countries greater geo-political importance). Europe was also criticised in Asia for not doing enough to help out. But there was a recognition that the crisis required American political and financial leadership. This was galling to many of the region's leaders who had made much of their self-reliance and had railed against what were perceived as US attempts to impose its political values and economic rules on the world as a whole.

The IMF's managing director Michel Camdessus, as part of his argument for funding from the US, claimed that the fund's disbursements in Asia were in America's interests. That was precisely the accusation levelled by nationalists in Asia: that structural reforms in their markets (in particular in banking) meant prising them open to the benefit of US companies. Few were as explicit as Mahathir in denouncing this 'new form of colonialism', but the resentment was shared by many ostensibly acquiescent Asian politicians.

This has to be seen in the context of the frequent skirmishes between the US and Asia over a range of issues. The countries of South-east Asia share, in varying degrees, worries about China's rising political and military might. Most, partly for that reason, welcome American involvement in the region's security. Yet, whenever ASEAN meetings bring all these countries together, the ASEAN countries routinely squabble with the Americans, and find China on their side.

Once economic recovery is underway, a backlash against US influence is quite possible. Indeed, by early 1998 it had, to a certain extent, already begun. It could be seen in Suharto's attempts to find a quick fix outside the scope of his IMF programme; in Mahathir's futile efforts to rewrite the rules of intra-regional trade; and, in a more benign form, in the nationalist (and economically insignificant) campaigns introduced in Malaysia, Thailand, Indonesia and, especially, South Korea, to mobilise donations from the public to mitigate the effects of the slump.

Where To Turn?

One cause of American dominance was the failure of the two regional giants – Japan and China – to provide local leadership. The refusal of Japan to provide a large fiscal stimulus to its own economy, which might then be in a position to provide a locomotive for regional recovery, was seen as central to the region's troubles. China , although in an extremely healthy external financial position, and a lender in the rescue packages, had enough economic worries of its own to be seen as potentially part of the problem rather than of the solution.

In early 1998, regional recovery hinged on the fate of Indonesia more than on any other single factor. If that country can negotiate its political shoals, then it is possible to envisage, after several years of economic hardship, a more open regional economy. If the IMF's prescriptions are followed through, Indonesia would have the most liberal economy in Asia. Its neighbours would be compelled to follow suit. If, however, there is a cataclysm and a period of prolonged political uncertainty, foreign investors would flee, and Indonesia might retreat behind the protective wall of capital controls that would be needed to make a currency board stick, and revert to old protectionist, nationalist trade policies. That, too, might have a regional knock-on effect. Even taking the optimistic scenario, it would be wrong to see the current period of US economic and political influence as permanent. The hurt inflicted by the events of recent months will lead regional leaders to look for ways of ensuring it does not happen again. They might include a search for greater diversity in the currencies of their trade, a more domestically and regionally based growth strategy and an attempt to build stronger regional defence mechanisms to counter international trends.

◆

Caspian Oil: Not The Great Game Revisited

The analogy between the strategic aspects of the development of energy in the Caspian region and the great game of nineteenth-century competition between Britain and Russia for influence over the region is often drawn. On further examination, however, while the label may be journalistically convenient, the parallels between the current strategic positions in the Caspian and their nineteenth-century counterparts are extremely weak. The great game was played out between two nations in the context of what amounted to a complete power vacuum within the region. Today, the number of external players is large, their aims far more complex than the rather black-and-white imperatives of the great game, and there is no longer such a convenient vacuum for the external players to fight over.

Perhaps the major surprise since the collapse of the Soviet Union has been the robustness of the new Caspian states: Azerbaijan, Armenia, Turkmenistan and Kazakhstan. Most early analyses placed great stress on the competition for influence between Turkey and Iran, assuming that the new states would simply divide into blocks determined by the interaction of various forces, particularly considerations of language groups and religion. Indeed, large sums of money were pumped into the area by concerned countries in explicit attempts to stop the formation of any area of direct Iranian influence. Other early analyses concentrated on the thesis of a fast return to complete hegemony by Russia. This view was reinforced by the fact that all the transport, communication and economic logistics of the area had been carefully designed over the long period of Soviet rule to run to the centre and not radially, in order to dampen any centrifugal forces and frustrate the viability of independence movements.

In the event, neither scenario has come true. The new states in many cases have been quick to exert independence and to forge distinct national identities. They have proved extremely adept at playing outside interests against each other, steering adroitly between Iran and Turkey, between the US and Russia, and between China and the West. For example, the composition of interests in Azeri oil consortia is not a direct result of economic forces, but the result of a very deliberate weighing-up of Azeri foreign policy interests.

To consider the strategic positions of the major external players involved in the pipeline politics of the region, it is important to put Caspian oil and gas within the context of the world market. In short, Caspian energy is much less important than many political analyses have implied. There are three main reasons why it would be unwise to overestimate the potential of this oil: scale, cost, and timing.

Big, But Not That Big

Much of the inflation of the real importance of Caspian oil has derived from US Department of Energy estimates, publicised by the State Department, which placed the potential ultimately recoverable reserves of oil in the region at around 200 billion barrels. By way of comparison, the original recoverable reserves of the two largest fields, found after 140 years of exploration in the US (the Prudhoe Bay and East Texas fields) amount to a combined total of 18bn barrels. The size of the largest oil field in the world (the Ghawar field in Saudi Arabia) is 80bn barrels. A giant field is usually defined as one with at least 500,000 barrels of recoverable oil, and only about 370 such fields have ever been discovered worldwide. The US State Department figures imply that the Caspian area contains the equivalent of

400 minimum-size giant fields, or 16 Prudhoe Bay fields, or two and a half Ghawars. All this is based on little more than pure speculation.

Within the oil industry, the 200bn barrels estimate is widely derided, but it is the figure which recurs frequently in journalistic and political analysis, and the figure which has stoked up the general and political interest in Caspian energy exploitation. The implication of the large estimates is that the Caspian is only a little smaller in scale as an oil province than Saudi Arabia. This leads, in its most extreme form, to the conclusion that the Caspian is in some way a potential substitute for Saudi Arabia and the Gulf as a source of world oil supplies, or will enable substantially less dependence on these regions, with all this implies for the direction of US foreign policy, in particular.

Despite their near universal quotation, the US State Department figures are generally perceived to be an order of magnitude away from reality. The current proved reserves of the Caspian area as of 1998 are estimated by the *Oil and Gas Journal* at about 8bn barrels, and by the BP *Statistical Review of World Energy* at about 16bn barrels. The consensus of oil industry forecasts of the ultimate total recoverable reserves (including discoveries not yet made) tend to lie in the range of between 25bn and 35bn barrels. The North Sea rather than Saudi Arabia is thus a better point of comparison. Given this size, the possibility of the Caspian serving as a major long term competitor and substitute for the Gulf evaporates. Instead of the 16% of world reserves the US State Department implies, the true figure for the Caspian is likely to be closer to 3%.

It Is Expensive Stuff

The economics of Caspian operations provides a second sobering proviso. Caspian oil is extremely costly, with the price driven up by the difficulties of moving equipment into the area, the expense of pipeline construction, and, most significantly, by the transit fees payable to other countries on potential pipeline routes. Since the Gulf Crisis of 1990–91, oil prices have moved in a range between $12 and $25 per barrel. At the high end of this range, Caspian oil would provide good rates of return. At the low end, it would provide extremely poor rates. Any sustained period of low prices would cause interest in the region to wane and development timetables to slow. At the low end, gas development would become completely unviable.

In early 1998, prices were at the low end of the range, and the profit, net of development costs, operating costs, pipeline construction costs and transit fees, for currently active or proposed projects in the Caspian were derisory. Given the choice, international oil companies would allocate very little capital to the Caspian, were the low-cost reserves of the Middle East available to them. In this context, the potential for opening up the Iraqi oil industry represents a major threat to the development of Caspian energy. Iraq has far more oil than the Caspian, production can be increased on a far faster timetable, and it would be at an extremely low cost. To give but one parameter, Iraqi oil goes through Turkey, with transit fees of less than $1 per barrel, while Kazakh oil coming through Russia currently pays about $6 per barrel. Given the number of major deals signed already in Iraq, and the avalanche of deals to be expected should international politics allow them to occur more openly, accelerated Iraqi development would make the Caspian rapidly appear to be a backwater of the international energy industry.

It Won't Be Available Soon

The question of the timing of incremental Caspian production represents the third main proviso in assessing its importance. In 1997, the Caspian area, including those parts in Russia and Iran, produced about 1.2 million barrels per day, primarily from Kazakhstan (0.55 mb/d) and Azerbaijan (0.2 mb/d). Based on the current timing of potential projects, the Oxford Institute for Energy Studies (OIES) calculates a very slow take-off in production. On the most optimistic (and probably over-optimistic) timetables, Azeri production will reach 0.4 mb/d by 2000, 0.8 mb/d by 2005, and 1.6 mb/d by 2010. Base case projections by the OIES, allowing for a realistic timetable, place production at slightly less than 1 mb/d in 2010. For Kazakhstan, the projections imply production of 0.8 mb/d in 2000, a high of 3.4 mb/d and a more realistic base case of 2 mb/d in 2010. Adding in other areas produces a base case figure for the whole Caspian region of 3.5 mb/d in 2010, equivalent in scale to Norway's production as of early 1998. The additional production of slightly less than 2.5 mb/d over 13 years could be matched by an unconstrained Iraq in perhaps three years.

China's Unique Position

This is not to dismiss the importance of the Caspian, merely to debunk some of the hyperbole that has been expressed following publication of the reserve figures the US State Department used, and to note that the politics of Caspian oil are considerably more interesting than its impact on the international energy market. Diminishing the importance of the region as an energy province does not detract from the significance of the complex strategic aims of the players involved.

There is probably only one external player, China, which views the issue primarily as one of energy. Its objectives are based on its domestic energy situation; its main aim is to get access to resources. Political influence in the Caspian is seen not as an objective in itself, but as a means of tapping into a new source of oil and gas. A secondary objective is to reduce the possibility that ethnic unrest in western China will receive more active external encouragement.

As China became a net energy importer in the early 1990s, the Chinese oil industry placed its major hope on the prospects for production in the

Tarim Basin in Xinjiang province in the far west of the country. Since its offshore reserves were not living up to expectations, the potential for the Tarim was talked-up both by the Chinese and by international companies. As the years went by, however, disillusionment set in. The international oil companies were given only very marginal acreage, and the prime acreage held by the onshore national oil company, the China National Petroleum Corporation (CNPC), proved extremely disappointing. It became clear to the state planners that the politically important goal of domestic self-sufficiency in oil was not attainable.

An expedient adjustment to the language helped solve the problem. Self sufficiency would now be defined to include oil produced by Chinese companies, even if that happened to occur outside China. Rather than relying solely on the international market, it became imperative for CNPC to become an international player. In 1997, China launched a diplomatic offensive in Central Asia, and simultaneously CNPC set out to look for deals in the Caspian, in Iraq, and as far afield as Venezuela. The most significant agreement was in Kazakhstan, where CNPC won the Aktyubinsk field, ahead of a number of distinctly surprised US concerns, in a deal worth some $4.3bn.

China brought two elements to the table for Kazakhstan. First, its interest offered the Kazakhs leverage to use in negotiations with other competitors. Secondly, the Chinese, with their plan to build an oil pipeline into China, promised a solution to the impasse over western exit routes. (See map, p. 240.) From a Western viewpoint, that pipeline makes little economic sense, but Chinese pipeline and steel economics are radically different. Steel is still part of the state sector in China and thus its price bears no resemblance to international prices.

The pipeline is also but one element of a series of possibilities under active consideration, the so-called 'energy Silk Road', which involves gas pipelines from Turkmenistan, from both eastern and western Siberia, and future oil transportation options out of the Middle East. Given the scale of its energy requirements, and the large and growing size of its foreign currency reserves, China's potential to eventually exploit the eastern route should not be underestimated. In the interim, the shock that the Aktyubinsk deal gave Western companies may be repeated elsewhere in the Caspian. A strategic diplomatic policy, primarily motivated by energy policy goals, is a powerful force for signing deals and mobilising resources. By contrast, the other external player governments mainly have a Caspian energy policy which is motivated by diplomatic policy goals.

Others In The Game

China is, then, a special case. Other external governments tend to have aims which are to some extent obstructionist, either of Caspian development *in toto* or of specific development options. In several cases, policy is being driven by compromises between often divergent internal lobbies and interests, producing policies that have not necessarily been consistent. The major example is US policy in the region, which arises from several concerns. In the starkest terms, Washington's aims are to contain Russian and Iranian influence and to protect the interests of US companies.

This relatively straightforward goal, however, is complicated by the concerns of a long list of powerful lobbies, of which three may be highlighted. First, predictably, the Israeli lobby is important in the context of the relationship with Iran. While many within the State Department see the Caspian as potentially allowing the start of a breakout of the impasse over policy on Iran and the contradictions imposed by the policy of dual-containment, there can be no doubting that the Israeli lobby has proved to be an impediment. From a purely strategic perspective, Iran is a natural ally for the US (and, for that matter, for Israel) within the region, for it can facilitate southern export options as an alternative to Russian control and it can provide the primary bulwark against Iraq. In this context the pragmatism shown by the US in removing objections to Turkmen gas moving through Iran is a hopeful sign, even though it is only a small item when put against the major choices that need to be made.

Another lobby, while less powerful, is still capable of creating roadblocks. The Armenian lobby has been very effective, in that Azerbaijan has been the only part of the former Soviet Union to face US sanctions. Assuming that the Nagorno Karabakh issue remains essentially unsolved, the Armenian lobby still has the power to disrupt planning for the area. It plays a large role in forcing all proposed southern routes for the bulk of potential incremental Azeri supplies, (the so-called 'late oil'), to skirt Armenia, either through Georgia and then on to Turkey, or through Iran and then on to Turkey. (See map, p. 238–9.) In terms of US interests alone, the politically impossible Baku–Armenia–Turkey route would have been best. It would give Armenia a flow of revenue through transit fees, would cut Iran out of a major part of the equation, and it would remove exposure to the varying instabilities of the situation in Georgia.

The third lobby that affects export routes is that group of interests, including both human rights and feminist movements, that in practice remove the option of going east out of Turkmenistan through Afghanistan, even though deals have been made with all the factions in Afghanistan. Unocal, the US company involved with this proposal, currently faces litigation in the US over its human rights record in its operations in Myanmar, and this only adds to the likelihood that it will be unable to force through the Afghan route.

The interests of the US oil companies involved in the Caspian region have not had major effects on anything more than commercial policy. The companies themselves are well aware of the marginal nature of the oil and gas operations, and have been influenced by the extreme exasperation and costs incurred by Chevron in the earliest of the new Caspian ventures in the Tengiz field in Kazakhstan. They are also publicly opposed to the policy of dual-containment (and are much more vocal in private). They want low-cost, diplomatically protected exit routes, and in particular they want rapprochement with Iran, in at least as far as energy operations are concerned.

Given the above concerns, together with the stark and almost brutal objectives of some influential members of Congress, like Senator Alfonse D'Amato, US policy is hostage to the jockeying between groups. In strategic terms, however the key trend is that the logic for reformulating policy towards Iran is growing stronger, while still subject to the constraints epitomised by the Israeli lobby and Senator D'Amato. In short, it is becoming ever clearer that, if it is to make an impact, the US will need to imbed its tactical policies toward the Caspian within a more consistent strategic policy towards the Middle East.

Russian interests are perhaps clearer, but still arise from a coalition of interests. Russia would ideally prefer there to be no oil or gas in the Caspian. Given that there is, Russia naturally wishes to exert control through northern export routes, obtain a resolution of the issue of Caspian Sea property rights (giving Russia an increased share of the spoils) and exert leverage at every stage of negotiations. In addition, Moscow views it as essential, particularly after the shock of the Chechen crisis, to contain any expansion of ethnic problems along its southern borders. In this context, the worst outcome for Russia is to be faced with a group of prosperous Caspian states heavily influenced by either Iran or the US. What sometimes appears to be an effort to achieve regional hegemony in the Caspian is actually a symbol of significant domestic objectives and fears.

If the wishes of US energy companies do not loom large in the formulation of US policy, the same can not be said of their Russian counterparts. They are the major source of cash flow for both the economic and the political system, and each is tied in with some political faction or interest. In general, their objectives of securing as much Caspian business and as much rent as possible out of their pipeline system, has tempered the Russian tendency towards obstruction. While Russian–Caspian policy rhetoric (and Western analyses of that rhetoric) are still expressed in terms of strategic concerns, implementing that policy has become increasingly pragmatic and driven by more commercial matters.

Iranian and Turkish concerns also involve questions of influence, but despite the old analysis of a supposed Iranian and Turkish great game, the stances of the two countries increasingly work incidentally in the other's interest. For example, a major uncertainty for northern pipeline exit routes to the Black Sea is the Turkish desire to limit traffic through the Bosphorus. While this has an environmental basis, it is also a major bargaining counter for the country whose overall strategic influence and importance was the most drastically reduced of any by the time of the Soviet Union's collapse. At the same time, however, it reinforces the desirability of southern routes through Iran. Likewise, Iran promotes Iranian routes as one measure that will help it both break out of international isolation and also gain influence in the Caspian, but it incidentally reinforces the case for routes terminating in Turkey. Like the Russians, both countries face a range of internal ethnic issues, including the Kurds in Turkey and the ethnic Azeris in Iran. Although these add some complications, the general thrust of both Iranian and Turkish policy has been to find solutions to the major Caspian export issues.

Politics Over Economics

It is clear that the major threats to Caspian energy development come more from economics than politics. Indeed, if the price of oil remains low for any prolonged period the viability of the oil and gas fields will become questionable. Yet, to a large extent the question of development is almost a side issue for all bar the new Caspian states themselves and for China. What is really being resolved in Caspian energy negotiations is the nature of those states and their outward orientation, together with the increasing tendency for Caspian energy issues to become a conduit through which elements of the general alignment of Middle East policies can be altered. To paraphrase a cliché, while the candle in this case may be not prove to be that valuable, the game itself, and the way that it is played, is extremely important.

---◆

Making A Bigger NATO

At a ministerial meeting in December 1996, NATO members finally took the decision they had been edging towards for some time, announcing that the first new members would be invited to join the Alliance at a summit meeting in Madrid in July 1997. In the intervening six months, NATO and its member states were faced with three major tasks. The first was to seek, if possible, an agreement with Russia. By making clear that the timetable for enlargement was now fixed and non-negotiable, NATO ministers forced Moscow to decide between either accommodating itself to this new reality or continuing in its policy of uncompromising rejection. US Deputy Secretary of State Strobe Talbot and NATO Secretary-General Javier Solana led the efforts to persuade Russia of the benefits of cooperation and to negotiate a mutually acceptable agreement. The second task was to reach an agreement between the existing member states over which countries should be invited to join the Alliance. Although the NATO bureaucracy had been engaged in intensified dialogues with aspirant members to determine their fitness for membership, it was generally understood that the final selection would be a political, rather than a purely military or technical, decision. The final task was to provide reassurance for those countries excluded from the first wave of members.

Assuaging Russian Concerns

Despite the dire warnings from Moscow of its implacable opposition to enlargement, there were hints during 1996 that the Russian government was contemplating a less confrontational stance. In July 1996, Foreign Minister Yevgeny Primakov singled out the 'nucleus that is absolutely unacceptable to Moscow – moving up NATO's infrastructure to our borders. On this basis, Russia is inviting NATO to conduct a dialogue'. This policy stance laid the foundations for Russia's subsequent negotiating position.

On the general principle of NATO enlargement, Russia remained steadfastly opposed. With mantra-like repetitiveness, Russian leaders continued to assert that enlargement is the greatest strategic error since the end of the Cold War. But Moscow indicated that it might be less concerned if NATO's infrastructure was not expanded, meaning, in particular, no deployment of nuclear weapons or foreign troops on the territory of the new members. However, the ambiguity of the term 'infrastructure' also suggested that Russia wanted to limit, if not exclude, the incorporation of the new members into the Alliance's integrated military structure. In this regard, there was a particular concern that the enlarged NATO should not inherit the military infrastructure built to contain the Warsaw Pact. The Russian government also insisted that any agreement with NATO had to take the form of a legally binding treaty. A further demand was that there should be an institutional NATO–Russia forum which would ensure that Russia was both consulted about, and was directly involved in, NATO decision-making.

There was initial scepticism in the West as to whether these Russian demands provided a productive basis for negotiations. Accepting Moscow's demands would give Russia an effective veto over NATO decision-making and NATO would be unable to fulfil its treaty obligation of collective defence for the prospective new members. To prevent such a clearly unacceptable outcome, NATO defined its own conditions for negotiations. It was agreed that concessions could be made to Russia to sweeten the pill of enlargement, but only if such concessions did not breach what Strobe Talbot called the 'five no's':

- no Russian expectation of a delay in the process of enlargement itself;

- no Russian veto over NATO enlargement decisions or over NATO internal matters;

- no exclusion of any state over the longer term from the process of enlargement;

- no second-class membership for the new members; and

- no interference in NATO decision-making.

Although Alliance negotiators remained uncompromising over these exclusions, the subsequent negotiations in early 1997 did ultimately result in a package of reassurances which broadly satisfied Moscow. The NATO–Russia Founding Act, signed in May 1997, assured Russia that NATO has 'no intention, no plan, and no reason' to deploy nuclear weapons on the territory of the new members and that 'in the current and foreseeable security environment' there would be no additional stationing of 'substantial' combat troops. A Russia–NATO Permanent Joint Council (PJC) was established, but its remit was restricted to matters of common security concern and did not extend to the internal affairs of either party and did not provide them with any right of veto over each others' actions.

The Founding Act was also augmented by a number of other commitments which were perceived favourably in Moscow. First, and most significantly, NATO proposed reforms to the Conventional Armed Forces in Europe (CFE) Treaty which reflected Russian demands to move from a bloc-to-bloc to national and territorial limits. A further advantage of these proposals was the implication that the prospective national limits on the new NATO members would be low enough to exclude, in the legally-binding manner demanded by Moscow, the permanent stationing of forces from other NATO countries. Second, at the US–Russia summit in Helsinki in March 1997, President Bill Clinton outlined the framework for a Strategic Arms Reduction Treaty (START) III which would not only further reduce nuclear arsenals but would also directly address some of the perceived disadvantages of the START II which continue to hinder the ratification of the treaty in the Russian Duma. US support for Russian admission to the Group of Eight Industrialised Nations, the World Trade Organisation and the Paris Club were the finishing touches to the overall package aimed at securing Moscow's agreement to enlargement.

After the signing of the Founding Act, Russia continued formally to oppose the principle of enlargement, but this opposition lacked its earlier venom or vigour and was increasingly replaced by a pragmatic acceptance of the new geo-strategic reality. The Russian leadership, however, indicated that this relaxation of tensions might only be temporary. Primakov made clear after the Madrid summit that Russian cooperation with NATO might be swiftly terminated if there were any consideration of NATO enlargement to countries of the former Soviet Union.

Will It Be Three Or Five?

Securing Russian acceptance of enlargement was generally considered the difficult task. The candidates selected for membership were widely presumed to be almost a foregone conclusion. Although NATO was engaged in intensified dialogues with 15 aspirant countries, it was an open secret that Poland, the Czech Republic and Hungary were the clear front-runners. They were recognised to be among the more politically and economically advanced countries of the region and their relative distance from Russia meant that admitting them to NATO would be less provocative. Moreover, both the US and Germany, which had been leading proponents of enlargement, indicated their support for restricting the first wave of members to these three countries.

As the Madrid summit approached, however, the lack of consensus became apparent. Various NATO states promoted their own preferred candidates, even though some were generally recognised to be non-starters. Denmark and Norway, for example, valiantly promoted the candidature of the Baltic states. More difficult to dismiss was the case for Slovenia which had made impressive economic and political progress and, being a small country, could relatively easily be absorbed within NATO. Supporters of Slovenia's entry (Italy, and for a time the United Kingdom) argued that the relatively small costs of its admission were outweighed by the benefits of exporting stability to the Balkan region and demonstrating NATO's interest in the security of South-east Europe.

Slovenia's candidature was considerably complicated by Romania's strong and sustained campaign for admission. Until late in 1996, Romania was generally thought to fail most of the political and economic conditions for membership. But, in the November 1996 elections, the entire neo-communist administration was thrown out, including the President, Ion Iliescu. In a flurry of diplomatic activity, friendship agreements were signed with Hungary and Ukraine. In similarly swift order, a structure of formal civilian control of the military was instituted. As the Romanian claim for membership turned into a massive publicity campaign, critical external support was provided by France. President Jacques Chirac, smarting under the new demands of cohabitation, adopted the role of Romania's advocate for two principal reasons: first, to demonstrate France's continuing independence from US hegemonic control of the Alliance and, second, to provide a geopolitical counter-weight to perceived German influence in Central and Eastern Europe. In addition, Italy, championing a 'Southern element' to the enlargement policy, also became a strong supporter of Romania's claims.

The US sought initially to maintain a dignified silence at the increasingly fractious political infighting. But it was finally provoked when there were hints in spring 1997 that Germany, out of deference to French concerns, might look favourably on Romania's accession. In May, US Secretary of State Madeleine Albright made clear that the US would only support three countries for NATO membership at this stage. This unilateral intervention was driven partly by a sense of frustration that the Europeans had broken the implicit agreement not to advance the cause of individual countries prior to the Madrid summit. There was also a perception in Washington that the Europeans were promoting a larger first wave precisely in order to exclude any future enlargement. The US administration believed that an initially limited first wave of accessions to NATO would provide a better set of conditions for ensuring that enlargement remained an open-ended process.

The peremptory manner in which Washington set down its position upset even the traditionally more Atlanticist members of the Alliance. This demonstration of US power within NATO was a reminder of the Europeans' relative impotence in security and defence matters. In the face of the US *démarche* there was little they could do, and the final decisions made at the Madrid summit in July 1997 essentially conformed to the policy preferences set out by the US administration. Prospective admissions to membership were limited to Poland, the Czech Republic and Hungary. At the same time, however, the final Declaration emphasised the open-ended nature of the enlargement process. It was confirmed that the process would be reviewed at the next meeting in 1999, with a fairly explicit indication that Romania and Slovenia would be first on the list of any further entrants. The concerns of the Baltic states were also addressed by the specific reference to the states in the Baltic region which are also aspiring members.

Including The Excluded

These specific references to the continuing salience of the membership applications of Romania, Slovenia and the Baltic states considerably assuaged their fears of being left within a strategic no-man's land between East and West. In January 1998, the US also signed a US–Baltic Charter as a further symbolic confirmation of its commitment to the security of those states and its support for their ambition to become NATO members. Madeleine Albright's visit to the Baltic states in the immediate aftermath of the Madrid summit, where she asserted that the US was determined to reverse the legacy of Yalta, was similarly taken in the Baltic capitals to be a reassuring gesture.

Providing reassurance to Ukraine, which as a neutral country had not formally applied for membership, required a rather different diplomatic initiative. At the Madrid summit, a NATO–Ukraine Charter was signed, instituting a distinctive partnership between the two parties. Although the Charter largely mirrors the NATO–Russia Founding Act, the nature of Ukraine's relations with NATO clearly differ from Russia's, given that Kiev generally has a favourable perception of the Alliance. Both sides know, though, that the Charter is more symbolic than substantive in its content; there is no provision for hard security guarantees. The symbolism was enough to reassure Ukrainian officials. They were particularly pleased by NATO's definition of Ukraine as an East-central European rather than an East European country, which they interpreted as a recognition of their Western rather than Eastern orientation.

Measures were also implemented in Madrid to ensure that the wider institutional structures connecting member states with partners should be strengthened and enhanced. The North Atlantic Cooperation Council was replaced by the Euro-Atlantic Partnership Council and an enhanced Partnership for Peace programme (PFP) was established. Particular attention was given to incorporating partners within the reformed military command structure. A decision was taken to establish PFP staff elements at strategic and regional levels of the structure and a report was commissioned on establishing such elements at the sub-regional level. Providing structures for greater political consultation with partners and for increasing their role in planning PFP programmes were also highlighted and it was agreed that the Planning and Review Process for Partners would become more like the NATO force planning process. The general objective behind all these measures was to make the difference between being a NATO member and a partner 'razor thin', in the words of then US Ambassador to NATO, Robert Hunter. Partners were also expected to benefit from the increased regionalisation of the NATO command structure.

Nevertheless, it is widely recognised that these NATO measures need to be supported by the development of the European Union, which is itself engaged in a parallel process of enlargement. At the Luxembourg summit in December 1997, the EU accepted the recommendations of the Agenda 2000 Report to select Estonia and Slovenia, as well as Poland, Hungary and the Czech Republic, for negotiations towards membership. The decision to include Estonia and Slovenia within the first wave of EU members has done much to reassure the other Baltic and South-east European states that the process of both NATO and EU enlargement remains an inclusive rather than a divisive process.

For NATO, events in 1997 indicated that the relationship between the US and its European allies was far from harmonious. Tensions in the transatlantic Alliance are likely to intensify in 1998 with, as in the past, the issue of burden-sharing assuming centre stage. At first it was thought that the distribution of the costs of enlargement would be the major source of tension. This concern has been considerably alleviated, however, by the calculation of the modest nature of these costs, which are estimated to require only a $1.5 billion increment to NATO common funds over ten years. With this in mind, the burden-sharing debate has been transferred to a more substantive set of issues. Now the focus, particularly for the US, is on whether NATO and the European allies are really committed to assuming a greater responsibility for out-of-area collective security operations.

Yet, Congratulations Are In Order

In addition to intra-NATO difficulties, a harmonious NATO–Russian relationship is far from assured. Russian officials remain suspicious of the Alliance's objectives and there is a widespread perception that the West has failed in the past to honour the agreements and commitments made concerning NATO enlargement. Russia will thus remain highly sensitive to perceived slights as the agreements of the Founding Act are realised in practice. Russian officials will undoubtedly seek to promote discussions within the PJC which NATO member states would prefer to exclude as relating to their internal affairs. There are also numerous ambiguities within the Founding Act, such as over what constitutes a substantial combat force, whose stationing should not be allowed, and NATO's insistence on the provision of adequate infrastructure in the new member states. These can easily become the source of divisions and argument. In addition, Moscow's clearly stated opposition to membership of former Soviet countries, most notably the Baltic states, remains a substantial stumbling block.

Despite these concerns, it must be said that management of the difficult process of NATO enlargement during 1997 was a diplomatic success. NATO is on the path towards admitting three new members into its fold and so far has done so without unduly alienating or undermining the sense of security of those not chosen. The events of the year appear also to provide clear evidence that the process of enlarging Western institutions does help to export stability to Central and Eastern Europe. Ukraine has been the most evident beneficiary of this. In May and June 1997, immediately prior to the NATO–Ukraine Charter, Kiev signed a statement of Mutual Reconciliation with Poland, a long-delayed Treaty on Friendship and Cooperation with Russia, and a Treaty of Good Neighbourly Relations with Romania.

Russia's willingness to confirm Ukraine's sovereignty and political independence was particularly notable. It was driven by a belated recognition that earlier threats against NATO had tilted Ukraine into a more pro-Western stance. Some more optimistic commentators have argued that this rapprochement with Ukraine, alongside other developments in 1997, represents a critical turning point in Russia's evolution away from its imperialist past. The signing of the Founding Act, the confirmation of Ukraine's sovereignty, the agreement with the Chechens and the relaxation of the union with Belarus are all, it is argued, indicative of a Russia finally comfortable with its more modest destiny as a nation-state rather than an empire.

◆

NATO'S New Command Structure

Among international organisations, NATO has been conspicuously successful. One major reason why the Alliance both survives and thrives is because it is the only international security organisation which is expressly organised to deal with the most demanding of security challenges, the threat of an external attack. The Alliance alone has a standing military structure which is capable of moving smoothly from giving advice in time of peace to action in time of crisis. As the deployment of NATO's Implementation Force (IFOR) to Bosnia in January 1995 showed, even that most Atlantic-sceptic of countries, France, accepted the utility of NATO's military structure and moved more closely to it for practical reasons, while drawing back from formal integration for political reasons. The fact that Russia too, alongside a host of other non-NATO countries, found a way of joining IFOR indicates that, as an instrument of military power, NATO's military structure is recognised for its efficiency and effectiveness. Those countries which committed their troops to NATO's command in Bosnia did so primarily because they believed that the Alliance provided the best means of implementing the peace accord's complexities, such as disarming combatants and separating factions. Also, if the peace did not hold, NATO could be relied upon to protect and extract their troops. In other words, for complex, risky operations, NATO's command structure, and the (mainly US) military power it provides, is the security instrument of choice for both NATO members and many non-members alike.

The agreement on a new command structure, reached by NATO's defence ministers in December 1997, is therefore more than a technical issue. It is a significant sign of the way the Alliance is changing. It says as much about its future direction as any decision taken in recent years, even that of enlargement. How NATO is organised will determine its security options in the new century in Europe and beyond. The realisation of a new structure will also reveal how committed its members are to maintaining the Alliance as the primary instrument for their security, since in Europe there are alternative and competing options. The seriousness with which European allies approach the building of a European Security and Defence Identity through the Western European Union (WEU), or pan-European security cooperation through the Organisation for Security and Cooperation in Europe (OSCE), can be judged not by their statements of intent but by the extent to which they continue to invest heavily in the organisation of NATO, and lightly in that of the other institutions.

Old NATO: Old Structure

To appreciate the extent to which a new NATO has emerged in the 1990s, it is necessary to understand the structure of the old NATO. Although the Alliance has always had political aims and political leadership through the North Atlantic Council, its uniqueness lies in its highly developed military structure. While politically NATO sought to create the conditions of peaceful relations in Europe, practically and militarily it saw its function as defending what it termed the North Atlantic area, that is the territory of its members, against a threat from the Soviet Union. From the early 1950s, the Alliance developed a system of command in peacetime whose purpose was the defence of the territory of its front-line members. This command structure, which used to be termed the Integrated Military Structure, was capable of rapidly generating mass military force from its members, deploying it to defend the territory of the most exposed allies (particularly West Germany after 1954) and commanding those forces in war.

The structure had a single purpose and a single capability: the simultaneous defence of the whole of the NATO front line against an attack from the Warsaw Pact on all fronts. Thus the structure was elaborate, but not complex. It was hierarchical – at the beginning of the 1990s there were three Major NATO Commanders (MNCs) at the top of the system: the Supreme Allied Commander (Europe) (SACEUR), whose role was to organise the defence of NATO Europe; the Supreme Allied Commander (Atlantic) (SACLANT), whose role was to ensure the reinforcement of Europe by sea, and defend the lines of communication between North America and Europe; and the Commander-in-Chief Channel, whose area of responsibility covered the North Sea and English Channel.

Until the agreement on a new command structure, there were still four layers of NATO headquarters and 65 static HQs, many multinational, in the military structure. The purpose of each was to command and direct the activities of the layer below, so that in the event of an attack by the Warsaw Pact, the Alliance could put up to 115 division-equivalents into the field, in addition to sea and air power, and command them as an integrated military force.

However urgent the Soviet threat may have appeared during the Cold War, NATO never had command or control of forces in peacetime, apart from a few symbolic standing naval forces and air-defence assets. To generate mass force, it depended, and still depends, on its members fulfilling their planning commitments. Those countries which are members of the military structure (all except France and Spain) indicate to NATO's military authorities, in an annual return, the levels of readiness of the forces they would make available to the Alliance if there were to be an attack on NATO territory. Some forces would be declared at A-1, indicating that they would be made available within 48 hours and fully trained and equipped to meet that time frame. At the other end of the spectrum, other units might be declared at D-4, indicating that it would be several months before they would be available and that they would need training and equipment before they could be effectively deployed.

Under the NATO system, each country remained responsible for the size, organisation and funding of its own forces. But to enable the obligations of Alliance membership to be fulfilled, the allies created a system of military command which influenced the development of national forces in peacetime and is in place and ready to command those forces in time of crisis or war. The system and command structure was thus based on voluntary commitments generated in the Cold War by the self-interest of the allies in having a counter-weight to the Soviet Union and the Warsaw Pact. The success of this system, despite the absence of compulsion, could be gauged by the fact that, until 1990, 90% of NATO's land forces were maintained at two days' readiness or less; 70% of combat aircraft and all surface-to-air missile units were at 12 hours' readiness; 75% of major warships were maintained at two days' readiness.

The most distinctive feature of the old Alliance structure was its unambiguous US leadership. Out of political choice, the European allies engaged the US in the defence of Western Europe; out of military necessity they submitted to US military leadership and the constant US pressure to submerge national defence policies into a multinational NATO. Inadvertently, by investing so much into the development of NATO's military structure, the allies were creating an instrument whose value would transcend the end of the East-West divide in Europe and serve a different purpose after the Cold War. The realisation that this was the case came only after a period of political uncertainty within the Alliance at the beginning of the 1990s.

New NATO: Transitional Structure

During the Cold War, NATO's command structure was focused exclusively on the requirements of a large East–West war, and more specifically on territorial defence, as far forward as possible, against an armoured attack by the Warsaw Pact. The major military headquarters were static, not mobile, and their communication systems, as well as logistic support were organised accordingly. The purpose of this structure was exclusively deterrence through credible defence. Under the strategy of 'flexible response', which guided the development of Alliance forces from 1967–90, the option for first, and possibly early, use of nuclear weapons was a central element.

NATO's new Strategic Concept, agreed in 1991, started the reform of the Alliance by broadening its political objectives. All the ideas which were later to dominate NATO's agenda, such as partnership with former adversaries, crisis management and building a European pillar within the Alliance, featured in the new Concept. But while the Alliance was bold politically, it hedged its bets militarily. Flexible Response was abandoned as a strategy, and nuclear deterrence downgraded: NATO reduced its ground-launched nuclear stockpile by 80% between 1991 and 1993. Nevertheless, the Soviet Union, which still then existed, explicitly remained the yardstick against which NATO's capabilities were to be measured. Though no longer an enemy, the Soviet Union was considered to be still the only power in Europe capable of threatening Alliance members, and it remained NATO policy to prepare for that eventuality, however remote it might seem.

So while a new NATO was being born with ambitions to transform the face of Europe through cooperation and partnership, the old NATO remained structurally and stubbornly in place – just in case Europe, and particularly Russia, did not develop according to plan. There was some streamlining in the command structure to reflect the major reductions in the size and readiness of Alliance forces. But this was more facelift than radical surgery. The number of major NATO Commanders was reduced from three to two. The British Commander in Chief Channel – always the least necessary – was disestablished, leaving the two major US Commands (SACEUR and SACLANT) intact.

Figure 1 'Streamlined Structure', 1991

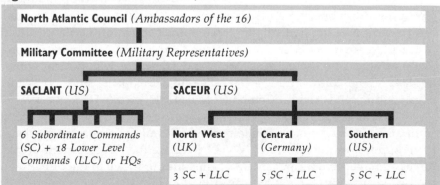

In total, the 1991 structure comprised 65 military headquarters of various sizes. The number and structure of these headquarters can only be explained in terms of Cold War priorities: such a command structure is required for the rapid generation of mass military force and an integrated (that is, multinational) defence against the East under single US military leadership. The command structure was not designed with problems in the Balkans in mind, and the 1991 streamlining did nothing to promote NATO's new role of crisis management announced in the new Strategic Concept. Continuity rather than change was the guiding principle.

Significantly, in view of the war in the Gulf in 1991 and the break-up of Yugoslavia, the streamlining failed to address the deficiencies in the southern region. The structure there was theoretical at best. Greece had withdrawn from the command structure in 1974, because of Turkey's intervention in Cyprus. The gap was never filled because of mutual blocking by Greece and Turkey.

NATO, even after the end of the Cold War, thus retained a heavy and static command structure for collective defence, while politically its leaders were looking to new roles for the organisation. The reduction of forces continued: by 1995 two-thirds of Allied land forces stationed in Germany in 1990 had left. Instead of 12 divisions (four US, three UK, three French, one Belgian and one Canadian) only one UK and two US full-sized divisions remain.

At the same time, a new force structure was being developed, made up of a limited number of reaction forces capable of deploying rapidly in a crisis, while the rest (main defence and reinforcement forces) would be available in a much longer time frame: 65% of the land force structure is now required to be ready only after 30 days notice. In theory, the development of rapid reaction forces created an Alliance more capable of crisis intervention than in the past. In practice, a new force structure was almost worthless without a suitable system of command. While the reaction forces trained for crises in the Balkans and the south, the command structure continued to be organised to face the traditional threat from the east.

The conservatism of the command structure became apparent from 1993 onward, when the Alliance began looking at plans for a possible NATO-led peace implementation force in the context of the Vance–Owen peace plan for Bosnia. It became clear to the then SACEUR, General Shalikashvili, that the existing command structure was not capable of deploying and commanding a peacekeeping force in Bosnia. Work then began within NATO on the Combined Joint Task Force (CJTF) concept.

A CJTF is a multinational, multi-service force specifically put together to fulfil a particular mission. While not new in military terms, it marked a turning point in the reform of NATO, because the development of such forces signalled the serious intent of the allies to move in the direction of their rhetoric. With the implementation of CJTF, crisis management is no longer a concept, but is becoming a mission that the Alliance is organising itself for.

It has nonetheless been a long haul. Creating the forces has not been the problem. As became clear when the Alliance put together and deployed IFOR, the military planning mechanism within NATO is highly efficient at matching the military resources of the allies, and now also the partners-for-peace, to the requirements of an Alliance plan. The key difficulty has been the relationship of CJTFs to the existing command structure.

Different allies have had different views of the concept from the outset. France, on the one hand, saw CJTFs as a completely new, alternative model of military cooperation for the Alliance, clearly different than the integrated command structure which France had left 30 years previously and did not want to rejoin. Moreover, with the idea of a future European defence policy in mind, several European allies regarded the concept as one which could serve the Alliance and the WEU in equal measure. The US, on the other hand, saw it as an additional feature of the structure, an improvement which would enable NATO to meet the full spectrum of its new missions just as effectively as the old structure had been able to fulfil its single mission of collective defence. The urgent need to put together the Bosnian peacekeeping force in December 1995 helped settle the argument. France and Spain, both outside the command structure, decided that they both could work within it if necessary.

Work within the Alliance has been focused on establishing CJTF Headquarters. The key difference between a CJTF HQ and the existing ones within the structure is that CJTF HQs are deployable to a theatre of operations; existing HQs are static, and are therefore limited in their flexibility. Against the background and experience of Bosnia, there is now agreement within the Alliance that three CJTF HQs should be established and attached to the command structure. While the creation of complete and fully manned CJTF HQs is recognised as both expensive and unnecessary in peacetime, the agreed concept provides for the establishment of HQ nuclei which could be rapidly brought up to strength when required.

Despite initial differences on the purpose of CJTFs, all allies are agreed that they will serve three purposes: to enable NATO to fulfil the full range of its missions (that is both peacekeeping and also possibly assisting in the defence of an ally against a limited attack); to allow NATO to support the WEU, thus implicitly making it unnecessary for the Europeans to develop their own separate force and command structures outside NATO; and finally to facilitate operations in which non-Alliance countries, and particularly partners-for-peace, can readily participate.

New NATO: New Structure

Until December 1997 and agreement on a new command structure, the Alliance can be considered as having gone through a transitional period. It

Figure 2 New Command Structure

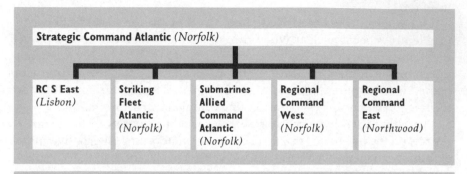

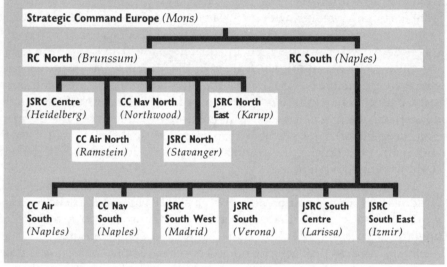

Key
RC Regional Command
CC Component Command (provides specialist air or naval support to other commands)
JSRC Joint Sub-Regional Command (Each of the JSRCs will have a focus to its mission. For instance, JSRC South is adjacent to areas of instability in the Balkans and North Africa. JSRC Centre, looking at potential risks from the direction of parts of the former Soviet Union, would focus on the collective defence mission.)

had defined a new, broader mandate for itself at the beginning of the decade. Since then, it had been struggling to organise itself to reflect that mandate. The NATO-led Implementation Force in Bosnia may have revalidated the Alliance as the pre-eminent security organisation in and for Europe, but its intervention was only possible because the military organisation had started to think through the problems two years earlier in the context of the on-and-off prospects of the Vance–Owen peace plan. A shorter notice, faster-burning crisis might have exposed the Alliance's deficiencies more clearly. At the time, NATO did not have the flexible command structures which would have allowed for a rapid response.

The new structure will take years to implement, but, when in place, there will be a further reduction in HQs from 65 at present to 20. The structure will be highly streamlined, and will eliminate a whole layer of military bureaucracy, comprising three levels in place of the current four.

This time round the restructuring does reflect a radical change in NATO's orientation. There will be three CJTF deployable HQs attached to the otherwise static system of command. SACLANT will have one seaborne CJTF capability against the event of a contingency operation that would require a predominantly naval command (a coastal evacuation or landing under fire). SACEUR will have two CJTF HQs available to him to plan for the rapid deployment of a task force and its command beyond the territorial limits of NATO's existing members in Europe.

There are two other equally radical, though less visible, features of the new structure. First is the increasingly close relationship and alignment between partners, under the Partnership for Peace programmes, with the command structure which will facilitate the rapid incorporation of partner forces into a NATO operation. The second is the capacity for the command structure to support European-led operations under the control of the WEU. Where the Alliance agreed, its assets and specifically a European-configured CJTF HQ could be made available to the WEU. The new command arrangements make it possible for the first time to disaggregate the Europeans without disintegrating the Alliance.

The success in arriving at an agreement on a command structure that reflects contemporary requirements should not mask the failures. The first and most obvious is the continued absence of France from the command structure, despite high hopes at the outset that a rapprochement was in sight. France argued for a European, on rotation, to take the Southern Region command. The US military rejected this as a challenge to it in a region where the US has major and growing interests and influence. The US press and Congress saw the French proposal as an attempt to gain control of the US Sixth Fleet, the major US commitment to NATO in the Southern Region. A number of imaginative ideas were floated, including a European deputy to the US Commander in the south in hopes of overcoming French objections, but in the end France could not accept joining a command structure which appeared to it to be more American than when France had left NATO in 1967.

A second drawback in the new structure is its fragmentation. It is less 'integrated', less capable of working coherently as a single unit, than it was. The disputes between Greece and Turkey on command arrangements in the Aegean were solved by the useful concept of a no-boundary command. Both Greece and Turkey get a sub-regional command each, but the relationship between them has to be determined only when a crisis requires it, the worst time to do so. The difference between Spain and Portugal over the command boundary in the Atlantic, which in Spanish eyes threatened to cut off responsibility for the defence of the Canary Islands from the defence of mainland Spain, was solved by linking the islands umbilically to Spain. The problem between the UK and Spain over Gibraltar was shelved by declassifying the British HQ there from being a NATO HQ. It continues to exist, but not for NATO command purposes.

Is The Debate Over?

Wiring diagrams can show only one part of the picture. Perhaps the more important point is that NATO has come through a decade of evolution with its most important asset, the command structure, reformed but still effective. What is less obvious is the number of additional attributes it has developed along the way. The new structure, with an inherent CJTF capability, will be able to respond to the full range of crises from peace-support operations, through defence of an individual ally or partner, up to and including collective defence, should that ever again become necessary. The role of the military in NATO, as nationally, is to advise their political authorities and to prepare to respond as directed to a developing threat. The range of options that NATO's military organisation can offer the North Atlantic Council has expanded considerably during the 1990s.

Similarly, the area to which the Alliance can deploy its forces has expanded to the point that the old debate about NATO 'out of area, or out of business' has lost much of its meaning. It is militarily conceivable, though as yet politically unthinkable, that NATO will extend its area of interest into North Africa and even the Middle East. New possibilities also lie in the ease with which Alliance structures can incorporate new members and also partners in its programmes. Even Russia can be accommodated within a military operation, as in Bosnia, in a way that flatters its self-esteem while increasing the leverage of the Alliance.

All this became possible through the 1990s, as NATO's military organisation and structure adapted to the new political and security situation. Before the 1990s, NATO had been a single-purpose, single-option organisation. Now it is far more complex, and far more flexible. If it has failed in any sense, it has been the continued absence of France from the command structure. Some progress has been made. France participates in CJTFs and every other major initiative of the new NATO and Spain is now a full member of the command structure.

Nevertheless, debate is bound to continue on the command structure. France continues to see the reforms as not going far enough in eliminating static structures that are better for territorial defence than for managing force projection. More radical reform would require NATO member states giving up status-enhancing command arrangements and will therefore be difficult to achieve. Yet the opportunity is there for progressive improvements on the command structure agreed after such tortuous negotiations in December 1997, especially if France and the US are able to carry on practical discussions about command arrangements outside the glare of publicity. It remains to be seen whether debate over the Alliance's new strategic concept will in turn open up a debate as to whether more tinkering with the command structure is needed to support the new concept effectively. In this sense, some may come to regret that the command structure was agreed before first agreeing the new strategic purposes to which NATO would be put.

◆

Nuclear Weapons: The Abolitionist Upsurge

Since the nuclear age began, there have been persistent efforts to put the nuclear genie back into the bottle. The International Pugwash Movement, for example, struggled for over 40 years to establish total abolition as the proper goal of nuclear arms control and to explore the conditions under which this goal could be achieved. But such voices were always very much in the minority, and far from the main lines of debate. During the Cold War the superpowers built vast nuclear arsenals, with tens of thousands of warheads on each side, and enshrined nuclear weapons at the centre of their defence strategies. The notion of eliminating nuclear weapons stirred neither wide interest nor support.

In the last few years, there has been a dramatic change and remarkable upsurge of interest in, and support for, abolishing nuclear weapons. This has included a series of high-profile studies as well as a widening web of prominent supporters of the idea. Since the mid-1990s there has been an unprecedented focus on nuclear abolition as a desirable policy objective. The view must now be taken seriously.

The Nuclear Abolition Debate

Several substantial studies in the early 1990s foreshadowed the more recent attention to nuclear abolition. These included the Palme Commission Report (1990); Stockholm International Peace Research Institute (SIPRI)'s *Security Without Nuclear Weapons?* (1992); and the Pugwash Movement's edited volume, *Eliminating Nuclear Weapons: Feasible?*

Desirable? (1993). But it is the series of projects, publications and developments since 1995 that have created a new visibility and momentum for the nuclear abolition movement.

NPT Review and Extension Conference

In retrospect, the current wave of interest in nuclear abolition is a by-product of the debate over whether and how to extend the Nuclear Non-Proliferation Treaty (NPT) and the diplomacy associated with the NPT Review and Extension Conference of April–May 1995. In advance of the conference, the nuclear powers were subjected to several years of mounting pressure from many quarters, including non-nuclear states as well as groups advocating arms control, to fulfil their obligation to pursue nuclear disarmament, as required by Article VI of the NPT. Many believed that the eagerness of the nuclear powers to achieve an indefinite extension of the NPT would give those advocating abolition leverage to force more progress in nuclear arms control and to compel more serious attention to the process of nuclear disarmament. On 25 April 1995, at the NPT Conference, an international network of non-governmental organisations adopted a statement, labelled Abolition 2000, urging the adoption by the year 2000 of a convention abolishing nuclear weapons. More importantly, the parties to the NPT accepted a set of 'Principles and Objectives for Nuclear Non-Proliferation and Disarmament' that explicitly reaffirmed the commitment of the nuclear powers to the objective of eliminating nuclear weapons. By the time the NPT was extended, abolition was in the air.

The Stimson Center Project

In January 1994, the Henry L. Stimson Center, Washington DC, launched a project drawing together a panel of distinguished scholars, government officials, and retired military officers to reconsider the role of nuclear weapons in US security policy. Its first report, *An Evolving US Nuclear Posture* (December 1995) called for a fundamental reappraisal of US nuclear policy, asserted the declining utility of nuclear weapons in advancing US interests, urged phased reductions in nuclear forces, and recommended an immediate commitment by the US to the long-term goal of eliminating nuclear weapons. Because this panel was chaired by former Supreme Allied Commander in Europe General (rtd) Andrew Goodpaster, and included such figures as the retired Generals William Burns and Charles Horner, and Ambassador Paul Nitze, among other prominent figures, its call for abolition was both visible and surprising. In March 1997, a second report, *An American Legacy: Building a Nuclear-Weapon-Free World*, endorsed by the same collection of individuals, was published. It reiterated even more resoundingly that a nuclear-weapon-free world is in the security interests of the US and that US policy-makers should undertake the vigorous pursuit of such a world. The Project on the Elimination of Weapons of Mass Destruction remains an ongoing activity of the Stimson Center.

Nobel Peace Prize

On 10 December 1995, the Nobel Peace Prize was awarded to the International Pugwash Movement and its president, Joseph Rotblat. Not only did the award put the global spotlight on the abolitionist message of Pugwash's leaders, but it opened the doors to senior government officials and the media. For many abolitionists associated with Pugwash, the award signalled that they had moved away from the fringes of the nuclear discourse.

The International Court of Justice Takes a Side

On 8 July 1996, the International Court of Justice (ICJ) ruled in an advisory opinion that the threat or use of nuclear weapons would be 'generally contrary to international law,' albeit permissible in extreme circumstances of self-defence. Although a somewhat ambiguous finding, it buoyed the abolitionist movement. Many concluded that the ICJ had effectively established the illegality of nuclear weapons. There was nothing at all ambiguous about the ICJ's unanimous judgement that 'there exists an obligation to pursue in good faith and to bring to a conclusion negotiations leading to nuclear disarmament'.

Canberra Commission on the Elimination of Nuclear Weapons

In November 1995, the Australian government convened an international commission formed with the specific mandate 'to propose practical steps towards a nuclear weapon-free world.' This was the first state-sponsored effort to explore in concrete terms the path to the total abolition of nuclear weapons, usually termed 'zero'. The Commission had seventeen distinguished, internationally diverse members. It included some stalwarts of the abolitionist community, such as Joseph Rotblat and Major Britt Theorin, but also included General Lee Butler, former commander of US strategic nuclear forces, Oxford Professor Robert O'Neill, former Director of the IISS, UK Field Marshall Lord Michael Carver of the UK, former French Prime Minister Michel Rocard, and former US Secretary of Defense Robert McNamara. This was not a group from whom consensus could be automatically assumed. Nevertheless, in August 1996, the Commission issued a consensus document (*Report of the Canberra Commission on the Elimination of Nuclear Weapons*) that rejected the utility of nuclear weapons and called for the five declared nuclear powers to make an 'unequivocal commitment' to eliminating them. Announced with fanfare, formally presented at the UN and promoted by a concerted publicity campaign, the Canberra Commission's report galvanised interest in nuclear abolition and has remained a focal point in subsequent discussion of this goal. Its political significance lay in demonstrating that support for abolition was widening beyond its usual constituency.

Statement on Nuclear Weapons by International Generals and Admirals

In a striking development spearheaded by General Andrew Goodpaster and General Lee Butler of the US, some 60 retired senior military officers from 17 countries issued a statement on 5 December 1996 proclaiming nuclear weapons a threat to humanity and calling for their elimination. As with the Canberra Commission, this was arresting because it suggested that support for abolition was spreading among groups that had not previously supported this view.

National Academy of Sciences Study on US Nuclear Weapons Policy

In 1996, the Committee on International Security and Arms Control (CISAC) of the US National Academy of Sciences undertook an assessment of US nuclear weapons policy under the direction of General William Burns. Its report, *The Future of US Nuclear Weapons Policy*, published in the autumn of 1997, called for fundamental changes in US policy and urged phased reductions in nuclear forces to much lower levels. It devoted a chapter to the prohibition of nuclear weapons, concluding that 'the time has also come to begin to devote serious attention to the prospects for prohibiting' nuclear arsenals.

Carnegie Commission on the Prevention of Deadly Conflict

A high-profile examination of whether and how conflict can be prevented in the modern world was undertaken under the auspices of the Carnegie Corporation of New York, a major US foundation, during the two years 1995–97. This involved another commission of 16 prominent individuals, headed by Carnegie's President, Dr David Hamburg, and former Secretary of State Cyrus Vance. Its report, *Preventing Deadly Conflict*, was issued in December 1997. Though not dealing specifically with nuclear issues, the report addressed them. It applauded the Canberra Commission report and commended it to national governments. It advocated the elimination of nuclear weapons as a significant element in building a system of conflict prevention and argued that, 'The only durably safe course is to work toward the elimination of nuclear weapons within a reasonable time frame.'

The Council on Foreign Relations Weighs In

In 1996, the Council on Foreign Relations formed a study group, the John McCloy Roundtable on the Elimination of Nuclear Weapons, chaired by General Larry Welch, the retired former Chief of Staff of the US Air Force. The main finding in his *Chairman's Report*, which was issued in January 1998, was that further reductions in nuclear weapons are both possible and desirable. This group did not endorse abolition nor come close to a consensus that a commitment to zero is presently desirable. The report did record, however, that there was some support in the group for these propositions. For long-standing abolitionists the striking fact was that the Council on Foreign Relations, the institution most strongly associated with the US foreign policy élite, would have such a study group in the first place. This symbolised the distance that the abolition idea has covered in recent years, from the outskirts of the nuclear debate to the mainstream of the US foreign policy establishment.

International Civilian Leaders Step Forward

Seeking to keep the idea of abolition in the spotlight and to preserve the sense of political momentum behind it, a pro-abolition group known as The State of the World Forum, headed by former US Senator Alan Cranston, organised a statement signed by 117 prominent leaders from 46 countries, including 47 past or present prime ministers. Notable among these were James Callaghan, Jimmy Carter, Mikhail Gorbachev and Helmut Schmidt. The statement, issued on 2 February 1998, acknowledges the Canberra Commission Report, invokes a moral imperative to eliminate nuclear weapons, urges the leaders of nuclear states to commit unambiguously to the goal of completely eliminating nuclear weapons, and presses for immediate efforts to marginalise nuclear weapons.

What Do They Advocate?

Each abolitionist project, report or statement has its distinctive attributes, conclusions or recommendations. But there is much common ground. Surprisingly, the rapid and complete elimination of nuclear weapons is not what is being advocated. Abolitionists overwhelmingly accept that this would take decades. Rather than demanding a move to zero in the near-term, what is being called for in most cases is a genuine commitment by the nuclear powers (particularly the US) to the long-term objective of achieving zero.

There are several paradoxes associated with this position. First, few of the studies or statements confidently assert that eliminating nuclear weapons is feasible. In fact, many of the recent studies concede that this is uncertain. The National Academy of Sciences study is representative when it says that 'It is not clear today how or when' abolition could be achieved. Similarly, the first Stimson report says that 'it is not clear whether elimination can ever be achieved'. Hence, most calls for a firm commitment to zero are coupled with appeals for serious exploration of the feasibility of this goal.

Second, there is not a common wisdom on what eliminating nuclear weapons actually means. Some see zero as a way to escape and repudiate nuclear deterrence; others think zero may be possible because nuclear deterrence – at least of the existential variety – will continue to function even if there are no nuclear weapons, because the knowledge of how to make them will always be there. Some see ballistic missile defences as an essential and necessary part of a world in which zero prevails; others regard such defences as anathema. Some believe that nuclear weapons should be entirely eradicated; others regard a zero world simply as an extreme state of de-alerting – that is the reduction of the readiness of deployed nuclear forces so that they cannot be used instantly or accidentally. Alternatively, they urge the retention of virtual nuclear arsenals that are not deployed but can be mobilised if necessary. Some suggest that a fundamental transformation of the international system is a prerequisite for, or a necessary accompaniment of, a world of zero; others believe zero can be approached without such a transformation. Some emphasise the role of international institutions in managing the transition to and maintenance of a regime of zero (perhaps via the international control of nuclear weapons); others see little or no role for such institutions. Those who wish to eradicate nuclear capabilities equate zero with making nuclear weapons illegitimate, and perhaps even criminal; those who envision a deterrence-based zero world accept the continuing legitimacy of nuclear weapons. In short, there is no single outcome which all abolitionists are aiming for. In its place, visions of zero abound, many of them mutually exclusive.

The third paradox is that all of the declared nuclear powers are already formally and legally committed to the long-term objective of eliminating nuclear weapons. This was most prominently accepted in the NPT and reaffirmed in the NPT Extension and Review Conference, and has been proclaimed in other contexts such as at the UN Special Sessions on disarmament. In the US case, at least, the aspiration has been reaffirmed by nearly every president since 1945. What is being expressed in the repeated calls for a new commitment to the zero objective is doubt about the sincerity of the previous expressions of support. Thus, the new abolitionist literature is full of calls for 'genuine, unequivocal' commitment to abolition, or for 'vigorous or determined pursuit' of the objective of eliminating nuclear weapons. Nevertheless, the US government's official position is that it is committed to the elimination of nuclear weapons as a long term objective.

Combining these points puts the abolitionists in the awkward position of strenuously advocating a policy objective whose attainment is, at best, decades into the future; whose feasibility is questioned even by its advocates; whose endpoints are unspecified; and which has already been accepted as the long term goal by the relevant states. This leads even some who strongly support nuclear arms control to suggest that the abolitionist upsurge is an unfortunate distraction, deflecting attention from the near-term arms control agenda to distant and possibly unattainable horizons.

Why Abolition Now?

How, then, can the level and intensity of the present abolitionist upsurge be explained? What is the point of pressing hard now for commitment to abolition decades hence? There is a range of answers to these questions, but underpinning them all is a sense that there currently exists a unique, and possibly fleeting, opportunity to advance the abolitionist agenda. The deep and bitter antagonisms of the Cold War made even discussion of abolition pointless. The removal of intense great power rivalry has altered the international political context in ways that enable the abolitionist agenda to be taken seriously. The drive to capitalise on this opening is made more urgent by the fear that new great power rivalry – possibly caused by the rise of China or the failure of democracy in Russia – may bring this propitious moment to an end. This concern is clear in a sentence that has been picked up from the Canberra Commission report, and echoed elsewhere in recent abolitionist literature: 'The opportunity now exists, *perhaps without precedent or recurrence*, to make a new and clear choice to enable the world to conduct its affairs without nuclear weapons.' The push for abolition is happening now because the opportunity exists now.

While this explains why the traditional abolitionists are pressing their cause, it does not explain why the circle of support for abolition is widening in the US and elsewhere. Here a second factor comes into play: with the end of the Cold War many have dramatically changed their view of the cost-benefit calculus associated with nuclear weapons. Increasingly it has become apparent that many, including, in particular, many military officers, accepted nuclear weapons only as a repugnant necessity of the Cold War. The abolitionist coalition is being swollen largely because of a growing belief that the costs, dangers and risks associated with nuclear weapons exceed any conceivable benefits they may have in the post-Cold War era.

Several other considerations reinforce the recent abolitionist upsurge. One is the proposition that abolition should be pressed now precisely because the road is long and full of obstacles. For example, it will take years, and perhaps decades, to build an adequate verification system to support a zero world. There is also the logic that it will never be possible to reach the end of this long road if the nuclear powers refuse to take the first step. A related point is that choices and actions taken today will shape options in the future; for many abolitionists, this leads inexorably to the conclusion that it is important to establish the long-term objective of eliminating nuclear weapons so that it influences present policies and priorities. Another immediate concern for abolitionists is sending a credible signal to the world that nuclear weapons are being devalued and (for some) becoming illegal; this signal is important, it is argued, in stemming the tide of nuclear proliferation.

The abolitionists also strongly believe that the nuclear powers must reflect the goal of elimination in their current behaviour if their commitment to the long-term objective of zero is to be truly credible, as it has not been in the past. This, in turn, will contribute to the achievement of various near- and medium-term objectives advocated by the abolitionist coalition. It is how the commitment to zero in the long run helps reduce the nuclear danger in the here and now.

The Road To Future Zero Starts In The Present

While the broad objective of those advocating the elimination of nuclear weapons is to secure meaningful commitments to zero from the nuclear powers, nearly every study or statement they have produced has its own list of immediate or transitional steps that would lead to this goal. There are important differences in the selection of steps and the priority given to them, but they have a number of common recommendations. Proponents of zero like much of the existing nuclear arms control agenda, including such items as the 1996 Comprehensive Test Ban Treaty and the fissile material cutoff, but several items are especially prominent in their thinking.

Support for nuclear reductions, beyond what is laid down in the Strategic Arms Reduction Treaty (START) process, is nearly universal in the abolitionist literature. Many studies offer their own schemes for phased reductions. Most offer ambitious medium-term targets for the reduction process. The National Academy of Sciences study, for example, recommends seeking a limit of 1,000 total warheads for the US and Russia combined, including not only strategic nuclear warheads but also tactical warheads and reserves as well; this would require going well beyond anything presently on the nuclear arms control agenda. For many, one of the powerful virtues of a focus on zero is that it strongly reinforces the logic of reductions.

De-alerting has attracted wide support from proponents of zero. This can involve not only standing down from hair-trigger alert procedures, but can progressively entail removing warheads from delivery systems, storing warheads at sites remote from their delivery vehicles, and even partial or full disassembly of warheads. As noted, some regard zero as the inevitable outcome of the logic of de-alerting; the time-to-launch in a world of zero will have been increased from minutes or hours to days, weeks or even months. The logic of de-alerting is so compatible with the abolitionist impulse that it figures prominently in much of the recent abolitionist literature. It headed the list of immediate steps recommended in the Canberra Commission report, for example. Similarly, the Carnegie Commission report advocated moving to zero deployed warheads; that is a long way from zero, but also a long way from the current force postures as well.

Most, although not all, proponents of zero are strong advocates of nuclear powers adopting the principle of no-first-use. This is important to abolitionists because implicit in first-use doctrines is an assumption of the wide utility of nuclear weapons (for example, to deter conventional attacks or chemical and biological threats). The abolitionist goal is to restrict the role and to limit, or deny, the utility of nuclear weapons. Once all nuclear states accept that the only purpose of nuclear weapons is to deter the use of nuclear weapons, it is only a short step to the conclusion, deeply compelling amongst abolitionists, that if no one had nuclear weapons, no one would need them. For example, the National Academy of Sciences study strongly recommends that nuclear deterrence should be confined to its 'core function', the deterrence of nuclear attack. And many advocates of nuclear elimination attach high priority to the achievement of an international convention on no-first-use.

Thus, while abolitionists have one eye fixed on the distant goal of zero, they also have an immediate agenda. One of their aspirations is to put the long-term goal at the service of this short-term agenda. Nuclear testing, first-use doctrines, or the persistence of large nuclear forces, are simply not compatible, in the abolitionist view, with a serious commitment to zero. If the nuclear powers want to make their promises to aim for the elimination of nuclear weapons credible, they will have to pursue something like the abolitionist agenda in the short-run. This, in turn, will move them down the road to abolition, and bring the elusive end a little bit closer. That is why the abolitionists believe that it is worth pressing now for genuine, unequivocal, vigorously pursued commitments to nuclear abolition.

The Game's Afoot

'We are at the beginning...of a great debate,' writes long-time abolitionist Jonathan Schell in his passionate article on going to zero in the 2 February 1998 issue of *The Nation*. If Schell is correct, in the coming years, the advocates of abolition will battle with their still dominant critics to influence and determine the policies and objectives of the nuclear powers. It is a sign of how far the abolitionists' cause has come, how much momentum their idea has gained, how much their ranks have swollen, that such a great debate may be in the offing. This contest is only possible because the abolitionists have gained a place in the mainstream of policy discussions. They have established a beachhead through the events and efforts of only the last few years.

Nevertheless, there is a substantial difference between this surge in interest and political momentum on the one hand, and the triumph of the abolitionist policy agenda, on the other. From the abolitionist perspective, some long-coveted breakthroughs, such as the signing of the CTBT, have occurred; some very heartening developments, such as the large reductions associated with the START process, are taking place; some significant elements of the abolitionist agenda, such as the fissile material cut-off, are very much on the current policy agenda; and considerable overlap has emerged between abolitionist recommendations and existing policy interests and priorities, particularly in the area of de-alerting. Much in the current scene should give heart to those interested in the elimination of nuclear weapons.

On the other hand, it is patently clear that the abolitionist influence does not dominate current policy. Ten years after the end of the Cold War and seven years after the collapse of the Soviet Union, the US and Russia still have not thousands, but tens of thousands of nuclear weapons. The only legally binding strategic arms control agreement, START I, has its origins not in the conducive post-Cold War period, but in the early years of the Reagan administration – a period known as the new Cold War. Subsequent agreements remain mired in political difficulties. And perhaps most disappointing from the abolitionist perspective, neither of the large nuclear powers shows any inclination to disavow nuclear weapons, nor do they display any acceptance of the proposition that nuclear weapons are of limited and declining utility.

On the contrary, Moscow appears to have re-embraced them as the solution to its new and unaccustomed numerical inferiority compared with its various potential rivals, east and west, and as a necessary compensation for its crumbling conventional military power. Russia's nuclear doctrine, insofar as it can be said to have a nuclear doctrine, appears to be the second coming of John Foster Dulles: massive retaliation is the order of the day. The US, too, has reaffirmed its commitment to nuclear deterrence, including the retention of ambiguity about whether or not it might use nuclear weapons in response to the threat of chemical or biological attack. This posture makes no-first-use incompatible with US nuclear policy and demonstrates that Washington continues to see utility in nuclear weapons beyond the core mission of deterring the nuclear forces of others. Thus, in some fundamentally important ways, present reality does not conform to abolitionist preferences.

The goal of the abolitionists in championing a great debate is to change these disappointing realities. Despite all the changes that have been noted, it is still too early to determine whether this great debate will happen, or whether the abolitionist moment will pass. And if this great debate happens, we cannot know how it will come out, given the mixed evidence so far and the powerful currents that still work against progress towards the elimination of nuclear weapons. But for the short run, the most important point may be that there is much on the existing arms control agenda that attracts the support of both the abolitionists and their critics; the agendas of these competing coalitions may not necessarily diverge radically for years or decades to come.

UNSCOM: Inspecting For Peace

Although Iraq's ability to threaten its neighbours was severely constrained by the Gulf War, the United Nations Security Council concluded that Iraq would only cease to be a threat when its weapons of mass destruction and ballistic missiles had been eliminated. The Council therefore placed two conditions on the cease-fire it offered to Iraq: that it should renounce such weapons and should accept a weapons inspection regime which could verify Iraq's adherence to the agreement.

The Security Council recognised that Iraq would make every effort to avoid compliance and armed itself with the means to deal with this. First, the cease-fire was a Security Council resolution adopted under Chapter VII of the UN Charter, thus making it an enforcement action. Second, the Council agreed that it would not lift the various sanctions against Iraq, imposed before the Gulf War, until Iraq was judged to be in full compliance with the cease-fire terms. It is in the context of these two points – Iraq's direct challenge to a fundamental objective of the cease-fire and the military options built into that cease-fire for ensuring compliance – that the crisis that started in November 1997 and led to the Annan-negotiated agreement in March 1998 should be seen.

Establishing the Inspectors

The UN Special Commission (UNSCOM) was established under section C of UN Security Council resolution 687 (1991). Its tasks, essentially, are:

- To carry out immediate on-site inspections of Iraq's biological, chemical and missile capabilities;

- To destroy, remove or render harmless all chemical and biological weapons and all stocks of agents and all related subsystems and components and all research, development and support and manufacturing facilities;

- To supervise Iraq's destruction of all its ballistic missiles with a range greater than 150 kilometres, and related major parts and repair and production facilities;

- To monitor and verify that Iraq does not use, develop, construct or acquire any of the items specified;

- To assist and cooperate with the Director-General of the International Atomic Energy Agency (IAEA), which has similar responsibilities with respect to Iraq's nuclear capabilities; and

- To designate, for inspection, sites which Iraq has not declared, as

UNSCOM deemed necessary to ensure the implementation of its mandate.

To achieve this mandate, UNSCOM has a powerful array of inspection rights, including the ability to designate any site for inspection or monitoring and to demand unrestricted freedom of access to those sites, without delay or hinderance, for its personnel and equipment. It is authorised to use whatever detection or monitoring technology it chooses, wherever it chooses, and to take samples for on-site or off-site analysis. It can interview any official on matters pertaining to its mandate and it can confiscate or copy any documentation it believes it will need. It is free to use aircraft anywhere in Iraq for transport, surveillance or monitoring, and to make unrestricted use of communications by satellite or other means.

Operating Procedures

Operating procedures were planned to be similar to, but much more stringent than, those developed for the Chemical Weapons Convention. The agreement specified that Iraq would make two sets of declarations. One would give a full account of the weapons, components and production facilities which it was now banned from holding; and the other would list its dual-purpose capabilities (those that could be used for proscribed purposes but which were, in fact, being used for purposes which were not proscribed). UNSCOM would then verify these declarations (through inspections and checking against information from other sources) and supervise the destruction of any banned items it found. UNSCOM was to establish mechanisms for monitoring Iraq's dual-purpose capabilities and imports into the future.

The original plan of action, written by the UN's Centre for Disarmament Affairs before UNSCOM was established, stated that the detection and destruction programme should be completed within 40 days. It was envisaged that, in time, the monitoring and verification activities of UNSCOM would be taken over by other multilateral arms control organisations – for example, the newly created Organisation for the Prohibition of Chemical Weapons in its area of expertise.

While the system was designed to be robust, it was hoped that Iraq would cooperate fully. It was reasonable to assume that Iraq would see it as in its interests to comply, given that sanctions would be maintained until it was deemed to be in full compliance with the weapons provisions of the cease-fire and given that the whole resolution was underwritten by the threat of resumed military action if Iraq breached its provisions.

In practice, Iraq's declarations of its banned weapons capabilities were clearly untruthful and it refused until October 1993 even to acknowledge its obligations with regard to the monitoring and verification of its dual-purpose capabilities. It made no declarations about such capabilities until January 1994. What it did declare in 1991 was that it had no nuclear weapons, no biological weapons, no 'superguns', some 11,000 chemical munitions using three agents (mustard, and the nerve agents sarin and tabun), and only 50 *Scud* missiles and 11 mobile launchers. Furthermore, it initially refused to countenance the destruction of any of its production facilities or raw materials.

UNSCOM, therefore, had to develop procedures for dealing with Iraq's obstruction and lies. While it had the right to inspect anywhere in Iraq, it needed to know where to inspect and what to look for. Its efforts fell into five categories as described below.

Information from Intelligence Agencies

In the summer of 1991, UNSCOM, as a completely new organisation with no intelligence resources of its own, had to rely on the international intelligence community and open sources to enable it to assess Iraq's declarations and, if judging them to be false or incomplete, to plan inspections to find the undeclared items. This process was spectacularly successful. In that summer alone, UNSCOM uncovered: Iraq's nuclear programmes; *Scud* missiles and mobile launchers that were not previously known about; 50 fixed launch pads; 23,000 additional chemical munitions; vast quantities of precursor chemicals for the production of chemical weapons and two sites associated with the biological programme (although Iraq even then refused to acknowledge their true use, claiming that one was part of a purely defensive programme and the other was used for the production of animal feed and bio-pesticides). Unfortunately, however, high grade intelligence was inevitably used up more quickly than new intelligence was generated, and UNSCOM had to find new ways to unearth undeclared weapons programmes and items.

Documentation

The nuclear revelations in the summer of 1991 came as a shock to Western intelligence agencies. Iraq had clearly made more progress towards a nuclear bomb than anyone had previously suspected. That posed a problem for UNSCOM and the IAEA: how could they account for all of Iraq's banned capabilities when their extent was not known? The answer was to find the archives relating to the weapons programmes. These would include progress reports, indicating the output and extent of the programmes, the people and organisations involved and the facilities used. The first inspection aimed at tracking down the documents was again spectacularly successful: the joint IAEA/UNSCOM team found documents reporting progress in the nuclear programme as of 1990. These documents were removed from the inspectors at gunpoint. The next day, the team found a large cache of documents (some 60,000) in another facility. The Iraqis again tried to retrieve the documents, but this time, on instructions from UNSCOM in New York, they refused to yield them. This triggered a four-day stand off during which the Iraqis refused to let the inspectors leave unless they yielded the documents and the inspectors refused to go without them. The crisis was resolved, under threat of renewed military action, and the inspectors left with the documents from the second site (but not the more valuable ones from the first site). The triumph was short-lived since Iraq immediately took action aimed at ensuring that inspectors would fail to find any of the archives again. A cover story was developed that the President had issued an order that all relevant documents were to be destroyed, and the sites were carefully cleared of any that existed.

Technology
In August 1991, in response to Iraq's efforts at concealment, the UN Security Council adopted a second resolution (707), explicitly permitting UNSCOM to use aerial surveillance over Iraq to help identify hidden capabilities. These flights started immediately, using a U-2 aircraft operated by the US with UNSCOM markings, control and commissioning. Other technologies, including ground-penetrating imaging, radiation detection and air and water sampling, were also used to identify suspicious sites.

Interviews
Iraq's measures to prevent UNSCOM accessing its archives left the Commission once again wondering how to get at the full extent of the programmes and capabilities it was charged with dismantling. It decided to try another tack – identifying the persons and organisations within Iraq involved in the programmes and, through interrogations, trying to break down Iraq's lies and wheedle out the truth – or at least gain clues as to where to inspect. This approach has proved invaluable.

Information from Suppliers
Suppliers were approached to find out exactly what was supplied, when, to which organisations, via what routes, and in what amounts. This route has provided key evidence allowing UNSCOM to identify both the organisations involved (and hence some of the personnel) and the quantities of raw materials imported. UNSCOM could also use the information to make its interviews more productive than would otherwise have been the case because Iraqi officials could be confronted with incontrovertible evidence which had to be explained. The officials then had to resort to ever more complex lies which were increasingly difficult to maintain in a consistent manner. Inconsistencies in the explanations were then used to justify further investigations.

Combining Techniques

The best example of how interviews with officials, together with information from suppliers, proved highly effective was the probe into the biological weapons programme. After Iraq had denied the full-scale offensive programme that initial inspections had strongly indicated, UNSCOM had little in the way of leads to get to the truth. The sites at Salman Pak and al Hakam had either been razed, or, if still standing, were so sanitised and reconfigured that further inspection yielded no additional evidence. UNSCOM was at an impasse. It did not accept the Iraqi story but was unable to break it down until, in October 1993, Iraq finally accepted its obligations in respect of its dual-purpose capabilities under the monitoring and verification regime. Under this procedure, Iraq was required to declare its dual-purpose capabilities and past, present and future imports of dual-use items. It took a year of inspections and badgering before Iraq made declarations that were anywhere near adequate. However, in November 1994, Iraq handed over new declarations. Some contained details not only of the imports, but also names and addresses of the suppliers. In the course of the next few months, UNSCOM wrote to the named suppliers and found that, in one instance, one company exported, not the kilogram quantities of complex growth media that Iraq had declared, but some 40 tonnes in one year to one of the Iraqi organisations known to have made purchases for the weapons of mass destruction programmes.

Other enquiries, coupled with new intelligence reports, indicated that Iraq had imported, via the same organisation, all the equipment and materials required for full-scale biological weapons production: biological feedstock, complex growth media, fermenters and fermenter controls, spray dryers, milling machines, grist for those machines and filling machines. It also had the munitions, from local sources or imported, into which to put the biological warfare agents. Armed with this evidence, backed up with copies of customs and invoicing documents, UNSCOM was able to confront the Iraqi officials who imported the items and demand full accounting for them. Often Iraq's accounts were nonsensical, and one by one they were disproved. UNSCOM informed the Security Council of the situation in its April 1995 report and diplomatic pressure helped force Iraq to admit to an offensive programme on 1 July 1995. Even so, Iraq threatened to stop cooperating with UNSCOM if its new revelations did not lead to the lifting of sanctions within two months.

UNSCOM welcomed the admission of an offensive programme, but insisted that the new Iraqi version did not account for the scale of the imports. Iraq claimed that its efforts to scale-up production had been unsuccessful (thus explaining the huge inefficiencies implied by the figures in its declarations) and that it had not moved on to the weaponisation stage.

UNSCOM's scepticism proved justified a month later when Iraq made new, but still incomplete and inaccurate, admissions about the extent of the programme. The day after the team left Baghdad, on 7 August 1995, the former head of Iraq's weapons programmes Lieutenant-General Husssein Kamal Hassan defected to Jordan. Iraq panicked, invited UNSCOM back to Baghdad and handed over some 600,000 documents, films, videos and microfiche which Iraq claimed had been 'found' at a farmhouse where the General allegedly had, unknown to his colleagues, hidden them. Among other things, these proved that the biological weapons programme was much larger than Iraq had 'fully disclosed' only days before, and included elements (substantial agent production, weaponisation and field testing of weapons) which had been brazenly denied until then. Even then, however, there were indications that the cache of documents had been selectively weeded, strongly suggesting that even they did not reveal the full extent of the programme.

Planning an Inspection, Recruiting Inspectors

There are two types of inspections: those for uncovering Iraq's hidden capabilities; and those to monitor its declared dual-purpose capabilities. Monitoring inspection teams are resident in Baghdad, and follow set procedures established for each given site, specifically adapted to the equipment, materials and capabilities of the site in question. The following description relates to inspections to track down banned items.

The starting point of any inspection is information and analysis. UNSCOM began by establishing a core of operational planners and an Information Assessment Unit (IAU) at its headquarters in New York. The role of the IAU is to compare Iraqi declarations with information from inspections and other sources, in order to determine which banned weapons Iraq may have been able to construct, and where it might have done so. This also helps to identify areas of capabilities that Iraq logically would have sought for the programmes and has not declared, hence identifying what to look for and what to interrogate Iraqi officials about.

Once the IAU identifies a site for inspection, or a line of questioning to put to particular Iraqi officials, the operational planners put together a team to conduct the inspection. There are no general purpose inspections. Each has a specific mission, and each inspection team is tailor-made for that mission. For example, an inspection of a chemical weapons storage site in the early stages would require:

- a chief inspector, who is familiar with Iraq's declarations, the team's inspection rights and the full inspection plan (which is kept to a need-to-know minimum for other members of the team), and with the stature and personality to cope with Iraqi objections and countermeasures. As time went on, it became more and more important that the chief inspector knew the history of Iraq's declarations and efforts to deceive the inspectors. This, naturally, narrowed the pool of inspectors who could serve effectively as chief inspector;

- an ordnance expert to deal with any unexploded bombs or mines at the site before the team entered;

- a structural engineer to declare any buildings safe to enter (or not);

- chemists, expert in chemical warfare agents;

- experts in munitions from various countries of origin, particularly chemical munitions;

- decontamination experts, in case of accidental spills;

- a medical team;

- a communicator and a report writer to record statements made by either side;

- a photographer to record equipment and materials found at sites; and

- additional inspectors to secure the site so that items could not be removed unobserved during the inspection.

An inspection of a chemical production facility would require a different set of inspectors, with the munitions experts replaced by chemical engineers expert in both chemical weapons production and civil chemical production (to help assess whether the equipment was being used for declared, non-proscribed use). There would also need to be experts in chemical production equipment (to assist in identifying the origins and capabilities of the equipment at the site).

Several inspectors from UNSCOM headquarters are normally part of each inspection to provide continuity. But the bulk of the inspectors are appointed specifically for each mission (although many serve on multiple inspections). They are made available with the help of their governments but are in the direct employ of the UN while on a mission. Without the ability to obtain these additional inspectors, UNSCOM would need a very much larger permanent staff, and would, in all likelihood, not have access to the same level of expertise. While governments are happy to help supply inspectors for one or two weeks at a time, they would not be able to lend the same inspectors to UNSCOM indefinitely as many of them come from private industry and academic institutions.

Inspection teams gather in UNSCOM's offices in Bahrain, where inspectors are trained and briefed on the inspection's mission on a need-to-know basis. They then fly into Iraq in UNSCOM's transport aircraft, returning to Bahrain after the inspection to write their report, which the chief inspector submits directly to the executive chairman of UNSCOM.

Why The Inspections Worry Saddam

UNSCOM has achieved more than any of its creators ever imagined – indeed, more than they imagined would be necessary. As President Bill Clinton has said, UNSCOM has destroyed more of Iraq's weapons of mass destruction than the entire Desert Storm operation.

Although impeded every step of the way, UNSCOM and the IAEA have nevertheless been able to uncover the following:

- various programmes to enrich uranium to weapons grade, laboratory-scale plutonium separation, and engineering facilities for the development of an implosion bomb for delivery by ground-to-ground missiles being developed indigenously by Iraq;

- chemical weapons production on a much larger scale than declared, involving five sites and encompassing additional, more modern agents not declared by Iraq (VX and possibly BZ); over 200,000 munitions; over 4,000 tonnes of bulk agent, and around 20,000 tonnes of precursor chemicals;

- three biological weapons programmes (lethal human pathogens, non-lethal human pathogens, and animal/plant pathogens). These programmes had successfully produced some 19,000 litres of botulinum toxin, 8,500 litres of anthrax and 2,200 litres of aflatoxin. Other agents being examined were gas gangrene, bubonic plague, ricin, haemorrhaging fever, rotaviruses, camel pox and plant rusts. Anthrax, botulinum toxin and aflatoxin had been weaponised in bombs and missile warheads and, to the surprise of Western experts, in tactical field artillery (122mm rockets and 155mm artillery rounds). Iraq had also experimented with a remotely piloted aircraft for spraying agent behind enemy lines;

- an assembled 'supergun' and parts and propellant for four others;

- more *Scuds* and launchers than Iraq initially declared, three development programmes for indigenous production of missiles with proscribed ranges (reverse-engineered *Scuds*, a two-stage missile capable of hitting Paris or Moscow, and a space re-entry vehicle system, capable of landing a one tonne warhead anywhere on the globe).

They have destroyed, or verified as destroyed:

- the production facilities and equipment associated with Iraq's various nuclear weapons programmes. The IAEA has removed from Iraq the plutonium, highly-enriched uranium and irradiated uranium it found;

- *Scud*-variant missiles; 19 mobile launchers; 76 chemical and 113 conventional warheads for *Scud*-variant missiles; 60 fixed launch pads in various states of readiness; production facilities; support apparatus (such as radar vehicles) and components;

- the 'supergun', its components and propellant;

- biological seed stocks and the biological weapons production plant at al Hakam;

- the chemical weapons production plant and equipment from the Muthanna State Enterprise and the facilities at Fallujah;

- some 480,000 litres of chemical warfare agents (mustard, sarin, tabun), 28,000 filled and 12,000 unfilled chemical munitions, and large quantities of 45 different precursor chemicals for the production of chemical warfare agents.

The achievements of UNSCOM and the IAEA go further than what they have uncovered and destroyed. So long as inspectors have unfettered access, Iraq will find it difficult to build and hide an operational capability for any of the banned weapons. The inspections, monitoring systems and import/export controls that have been put in place make it very much more difficult for Iraq to import banned items or dual-purpose items for clandestine activities. In addition, the longer Iraq is forced to keep any banned items in less than optimal storage conditions to hide them from the inspectors, the more these items will degrade and become less of a threat.

Does UNSCOM Have A Future?

Iraq has never cooperated with UNSCOM. At best, it has allowed the inspectors to roam around the country freely until they arrive at the gates of an establishment where illegal items are hidden. At this point the inspectors have been blocked until diplomatic demands and military threats resolved the issue. Iraq has never fully and honestly declared its holdings of banned or controlled items, the full extent of its past weapons of mass destruction programmes, nor the extent of its research successes. In short, the relationship between Iraq and UNSCOM has always been adversarial, and any UNSCOM success has come about in spite of the Iraqi regime.

Less obviously, UNSCOM's success was a direct product of the unstinting support of the Security Council. The Council's willingness to back UNSCOM unreservedly, and to underwrite its mandate with threats of resumed force, enabled UNSCOM to achieve great things. Without that support, it is most unlikely that Iraq's biological weapons programmes would have been unmasked. It took four years of seemingly unproductive effort to do so. Had the Council wavered during those four long years, pressures to accept Iraqi protestations that it had no programmes might have prevailed. Iraqi insinuations that UNSCOM was merely exhibiting bias by insisting on pursuing the biological investigations would have stuck and Saddam would have got away with it.

From 1996, however, the absolute support of the Security Council began to come into question. During the spring it failed on a number of occasions to give the inspectors the backing they needed when they met increasing obstruction of their efforts to inspect Special Republican Guard sites. The denouement of the 1997/1998 crisis over inspection bodes ill for the inspection efforts. Unless it is quickly, and positively, clarified through actions on the ground, the deal which UN Secretary-General Kofi Annan struck with the Iraqi leader opens up the possibility that UNSCOM's authority will be progressively diminished and with it the effectiveness of the inspection regime will be lost.

The agreement poses that threat because it seems to accept Iraq's allegations that UNSCOM inspectors are out of control, and this might end the unstinting Council support on which UNSCOM relies. If not properly implemented, it could interfere with the direct line of command from the Security Council to UNSCOM, thereby creating room for Iraq to play one party (the Secretary-General and his staff) off against another (UNSCOM). It could lead to a dilution of the clear language of resolutions 687, 707 and 715 with vague wording about inspections of the Presidential Sites being conducted under 'specific detailed procedures which will be developed', without stating who shall agree those procedures and thus opening the way for Iraq to state that it must be party to that agreement. It adds a Special Group of senior diplomats to the teams for inspection of sensitive sites. These two roles, inspector and national diplomat, are incompatible. The mere presence of the diplomats may allow Iraq to delay access to any site on the pretext that UNSCOM's proposed inspection methods are unacceptable and that Iraq needs to negotiate acceptable methods with the diplomatic element of the inspection group. In particular, it could remove the element of surprise which has been important in the success of many inspections.

The cumulative effect of these changes will begin to erode that other vital element of UNSCOM's success – its culture, which, in contrast with the consensus and permission-seeking culture of the humanitarian wings of

Figure 3 Summary of the 23 February 1998 Memorandum of
Understanding between the United Nations and the Republic of Iraq

• The Government of Iraq reconfirms its acceptance of all relevant
resolutions of the Security Council ... [and] further reiterates its undertaking to cooperate fully with UNSCOM and the IAEA.

• The United Nations reiterates the commitment of all Member States to
respect the sovereignty and territorial integrity of Iraq.

• The UN and Iraq agree that the following special procedures shall apply to
the performance of the tasks mandated at the eight Presidential Sites in Iraq:

  a) A Special Group shall be established [comprising] senior diplomats
  appointed by the Secretary-General and experts drawn from UNSCOM
  and IAEA, headed by a Commissioner appointed by the Secretary-
  General.

  b) In carrying out its work, the Special Group shall operate under the
  established procedures of UNSCOM and IAEA, and specific detailed
  procedures which will be developed given the special nature of the
  Presidential Sites.

  c) The report of the Special Group on its activities and findings shall be
  submitted by the Executive Chairman of UNSCOM to the Security
  Council through the Secretary-General.

• The UN and Iraq further agree that all other areas, facilities, equipment,
records and means of transportation shall be subject to UNSCOM
procedures hitherto established.

• Noting the progress achieved by UNSCOM in various disarmament
areas, and the need to intensify efforts in order to complete its mandate, the
UN and Iraq agree to improve cooperation, and efficiency ... to enable
UNSCOM to report to the Council expeditiously under paragraph 22 of
resolution 687 (1991). To achieve this goal, the Government of Iraq and
UNSCOM will implement the recommendations directed at them as
contained in the report of the emergency session of UNSCOM held on 21
November 1997.

• The lifting of sanctions is obviously of paramount importance to the
people and Government of Iraq and the Secretary-General undertook to
bring this matter to the full attention of the members of the Security
Council.

the UN, is and has to be impervious to the non-legitimate complaints of the host country. Unless there are members on the Security Council, like the US and the UK, prepared to react strongly to the inevitable Iraqi obstructions and to take action to see them removed, the Annan–Iraq agreement could cast a long shadow over the effectiveness of UNSCOM and the inspection regime. Provided the political resolve remains to overcome these obstructions, UNSCOM and the IAEA could still achieve their objectives.

♦

Quiet Reigns Over The Arms Control Agenda

In comparison with the heady days of 1996, with the passage of the Comprehensive Test Ban Treaty (CTBT) and the extension of the Nuclear Non Proliferation Treaty (NPT), 1997 was a year of rest for arms control practitioners. In this quiet year, the three big arms control events were not very exciting. The Chemical Weapons Convention (CWC) entered into force; agreement was reached on improving the International Atomic Energy Agency (IAEA)'s system for safeguarding nuclear facilities and materiel; and the negotiation of the Ottawa Treaty on anti-personnel landmines (APL) was successfully concluded in December. However, the Conference on Disarmament (CD) in Geneva, predictably, did not manage to agree even a programme of work, and the highlight of the forum was the appointment of a few special coordinators tasked with seeking the opinions of delegations. The Biological Weapons Convention (BWC) continued its ad hoc committee process on negotiating a compliance mechanism and establishing a rolling text (square brackets are still tortuously being removed). The CTBT Organisation Preparatory Commission (PrepCom) was established in Vienna and the first preparatory committee for the enhanced review mechanism of the NPT took place with few hitches but also few decisions and much frustration.

The UN Register on Conventional Arms received reports from its usual list of adherents declaring their transfers, and the 33 signatories of the Wassenaar Arrangement met and agreed to support a small arms moratorium in West Africa. The UN Expert Group on Small Arms completed its first report which made some practical suggestions for controls on the trade in light weapons. Follow-up studies in 1998 and 1999 will concentrate on the flow of ammunition and on possible action. The Joint Consultative Group for the Conventional Armed Forces in Europe (CFE) Treaty agreed the Basic Elements document in July and began work on adapting the Treaty in September. For most of the world's arms control endeavours, 1997 was very much a year of business as usual.

Ongoing Efforts To Control Nuclear Weapons

Phase one of the Strategic Arms Reduction Treaty (START I) was successfully completed by its third anniversary, with the US and Russia reducing their arsenals well below the agreed limits set for this stage. There were a total of 57 inspections under START I, 31 in the Commonwealth of Independent States (CIS) and 26 in the US. START II, however, is a different story. In March 1997, Presidents Clinton and Yeltsin met in Helsinki and forged a new deal over START II in an effort to encourage the Russian Duma to ratify it. It was a ploy that has yet to work. The agreement extended the deadline for the elimination of delivery vehicles under the Treaty to 31 December 2007, and agreed to begin negotiations on a START III deal as soon as START II enters into force. It was agreed that under START III each party would be permitted aggregate levels of 2,000–2,500 strategic nuclear warheads and that there would be transparency measures for warhead inventories and destruction. Delivery vehicles to be eliminated under START II should be de-activated by 31 December 2003, originally the deadline for elimination.

There are many and varied reasons for the delay in Russia's ratification of START II. The main ones include the feeling in Russia that the deal is not advantageous; anger and concern over NATO expansion; a change in Russian military doctrine which relies more heavily on nuclear weapons for defence against external threats because of the country's weak conventional forces; internal political debates in which the Treaty has been demonised and concerns over the costs of implementation. It is hard to see

Figure 4 The Wassenaar Arrangement

Member Countries (as at 1 August 1997)

Argentina	Denmark	Italy	Portugal	Switzerland
Australia	Finland	Japan	Romania	Turkey
Austria	France	Luxembourg	Russia	Ukraine
Belgium	Germany	Netherlands	Slovakia	UK
Bulgaria	Greece	New Zealand	South Korea	US
Canada	Hungary	Norway	Spain	
Czech Republic	Ireland	Poland	Sweden	

Established in July 1996, the Wassenaar Arrangement is a voluntary system to coordinate national controls on exporting conventional arms and dual-use technologies by promoting information exchange through a consultative forum.

how all of these concerns are being adequately addressed in Russia. There was some progress towards ratification, in that high ranking officials and politicians, including Yeltsin and Foreign Minister Yevgeny Primakov, made public statements in support of the Treaty and the government established a team of people to handle the ratification process. There was some talk that the Duma would ratify the Treaty in mid-1998; given that each of the last few years has begun with such optimism, it may be advisable to keep expectations in check and hope for a pleasant surprise.

The first NPT PrepCom meeting under the enhanced review process agreed at the 1995 NPT Review and Extension Conference, was held in New York in April 1997. Under a new 'cluster structure' for discussions – including nuclear disarmament, safeguard controls, nuclear-weapon-free zones and peaceful uses of nuclear energy – PrepCom issued a final report. The document recommended that the second PrepCom (to be held in Geneva in April–May 1998) should use the same cluster format and also address security assurances, the Middle East and fissile material controls. A working paper by the Chairman, annexed to the report, outlined the few areas where there was general agreement. There was a sense that this meeting had gone very smoothly, but little had been achieved in the wider sense of non-proliferation.

In May 1997, after years of study and negotiation, the IAEA was granted permission to adopt an enhanced safeguards agreement. The study was dubbed '93+2' (because it began in 1993 and finished in 1995), and although the adopted agreement is called the Additional Protocol, the arithmetic name has stuck. The new safeguards measures draw heavily on the IAEA's experiences in Iraq, North Korea and South Africa. The emphasis under the Protocol moves away from simply looking for diversions of discrete amounts of fissile material at a given site, but takes into account the possibility of diversion and proliferation within a whole country. Much more information has to be provided by the particular state to the IAEA, and there are new short-notice inspections and environmental sampling, allowing minute traces of material to be detected at places far from a site of interest. The Protocol gives the IAEA new measures, making it somewhat more difficult for a would-be proliferator to hide illegal diversions of fissionable material. States with nuclear ambitions would no longer be able to assume that it was going to be easy to evade detection by the IAEA. Indeed, the reverse is now true.

There was much debate over the additional burden that the new Protocol would place on states with large nuclear industries. In order to assist them, the nuclear weapon states will also participate in the Additional Protocol, thus creating a basis for an equality of misery for all concerned. The nuclear weapon states are not subject to all of the Protocol's provisions (it is, after all, designed to catch new proliferators), but they have opened up much of their non-military industry as a gesture of goodwill. The Protocol is voluntary and there are very few states that have formally adopted it. There will, therefore, be a long period during which the IAEA will operate under both safeguard systems. This will cause some confusion, as inspectors are being trained in the new approach and yet most of the inspections that take place over the next few years will be carried out under the old regime. There is also disquiet that only a small number of countries will sign the Protocol and that the new measures will not apply to states of most concern. Political pressure will be the only form of leverage to encourage full participation. As a result, there could be difficult times ahead.

Controlling Other Unpleasant Weapons

The CWC, negotiated in the CD and opened for signature in 1993, finally came into force on 29 April 1997, 180 days after Hungary became the 65th state to ratify. In the final few days before the CWC came into force, there was an undignified scramble to ratify the Convention in the US. The Senate eventually voted in favour of the Convention on 24 April, but it attached 28 conditions to its resolution. Had this set of conditions been attached to the ratification decision of any other country, particularly one which is on the US' list of rogue states, it would have caused an uproar. For example, the conditions prohibit any samples collected from inspections in the US to be analysed at a foreign laboratory; require that US research and development on improving verification procedures for the CWC be subject to cost-sharing between the US and the Hague-based Organisation for the Prohibition of Chemical Weapons (OPCW); demand voting rights in the OPCW even if the full amount of funds due is not paid; and require that the President certify annually that the CWC's limits do not significantly harm US commercial interests.

Russia missed the entry into force deadline but it did manage to ratify the CWC in November. By the end of 1997, 106 of the 169 signatory states had ratified the Convention. After the Treaty had become effective, the OPCW began the process of inspections and held the first Conference of State Parties. In a surprise move, Iran ratified the Convention and India declared that its stocks and production facilities were now subject to inspection. The international community was surprised by the number of countries that admitted to having chemical weapon programmes. China, India and South Korea were the high profile cases. Perhaps the most significant ratification was that of China which now, for the first time, accepted intrusive inspections on its own territory. A number of others also made initial declarations. Despite a serious cash shortage in the OPCW, inspections were begun in a large number of states in order to verify their initial declarations.

Probably the most significant disarmament event of 1997 was the establishment of a new partnership between governments and disarmament non-governmental organisations (NGOs) and international organisations (IOs) concerned with development, humanitarian and disarmament issues. The Convention on the Prohibition of the Use, Stockpiling, Production and Transfer of Anti-Personnel Mines and on their Destruction was the result of a partnership between concerned states (notably Canada and Norway) and a broad spectrum coalition of IOs and NGOs. Attempts to negotiate a treaty in the CD had stalled and a group of like minded states went ahead with separate negotiations in Ottawa and Oslo. Public opinion was strongly behind the ban, particularly as a result of publicity campaigns on the issue. The death of Diana, Princess of Wales, a highly visible advocate of a ban on anti-personnel landmines (APLs) served to highlight the problem even more.

The treaty text was adopted in Oslo in September and, in December, the Convention was signed at a large ceremony in Ottawa. The Convention is as much noted for those states which have not signed it as for those which have. Notably outside the treaty are China, Russia and the US. Although a large number of other states have indicated that they will not be signing the Convention, they are clearly beginning to feel the pressure created by the effective campaigns against the production and use of such mines. The Convention's supporters see the treaty as establishing a norm. They hope that international public opinion will eventually put enough pressure on those holding back to make them sign the document. There have been attempts at the Conference on Disarmament in Geneva to negotiate a ban on APL transfers. This would include many of the states that are not party to the Ottawa Treaty, but there is opposition to this move. There is deep concern that negotiating such a half-measure would undermine the moral high ground now occupied by the Ottawa Convention.

A New Look Is Required

Since the heady gains of 1996, disarmament and arms control is now in a period of consolidation and implementation. Having dismally failed to agree a programme of work in 1997, the CD, on 26 March 1998, did so for the coming year. Even so, that in itself will amount to little in the way of progress for it failed to appoint the personnel necessary to run it. The process at the CD is in a stalemate over the issue of nuclear disarmament and, until some compromises can be made between the hardline positions on both sides, the best efforts of the moderates will create little movement. Disagreement on nuclear disarmament is blocking all progress in multilateral disarmament forums. Indeed, the main lesson from the Ottawa Process may well be that, to conclude any disarmament treaty over the next few years, groups of like minded states will have to opt out of multilateral structures and make their own agreements. These can then be used to set future norms and standards for the rest of the world. Unless the whole security and disarmament agenda is restructured and rethought, Ottawa may well be the shape of things to come.

The Americas

The United States is generally thought a Puritan country, whose people frown on the usual weaknesses of the flesh, particularly those of their elected leaders. Yet, the sexual peccadilloes alleged against President Bill Clinton do not seem to have dented his popularity. His opinion poll ratings in the third week of March 1998 were at their highest levels ever. What appeared to be most important to Americans was that their country was at peace and was prosperous. The extraordinary economic boom, now entering its seventh year, had brought the lowest level of unemployment in 24 years and the lowest inflation rate for 30 years. There was no federal deficit for the first time in 30 years, and crime rates were falling sharply all across the country.

While the President had little to worry about on the domestic front, developments in the international arena brought new problems. The unexpected eruption of financial chaos in Asia and yet another challenge from Saddam Hussein in Iraq tested his ability to take the lead in solving, or trying to solve, international conundrums. Issues in China, Iran and Cuba all underlined how difficult it is for the US system to adjust its entrenched policies to accommodate rapid changes abroad.

The comfortable good times in the US stood in sharp contrast to the problems faced by its neighbours to the south. There was a surge in organised crime, corruption and insecurity throughout Latin America. The fledgling democracies in the region all survived, but the difficulty many were having in keeping their military forces under firm civil control showed just how fragile they are. Drug trafficking continued to be an intractable problem in many Latin American states. The few successful efforts to damp it down in one country only diverted the problem elsewhere. Nevertheless, the increasing participation by political parties and civil institutions in the democratic process, and the increasing involvement of Latin America in the international community, were positive and encouraging factors.

◆

The United States: Sitting Pretty

As President Bill Clinton took the oath of office on 20 January 1997 for his second term, the US basked in an extraordinary glow of good fortune. The economy was roaring along at some of the highest levels of growth and employment since the 1970s. The happy coincidence of such growth – in stark contrast to the economic malaise in Europe and the developing difficulties in Asia – and Clinton's successful re-election confirmed for the administration the basic soundness of the policies it had followed in 1995 and 1996. As a result, it advanced few major new domestic initiatives in 1997. Instead, it intensified the efforts begun in the first term, such as seeking a balanced budget agreement and gradually expanding health care coverage.

It was no different with foreign policy. For most of 1997, the administration concentrated on building on the agenda followed in 1996, championing an enlargement of NATO, maintaining a US presence in Bosnia, struggling to find peace in the Middle East, and improving relations with China. But the focus shifted to the Gulf at the end of 1997, as the administration found it necessary to rebuild its strength in the region dramatically to force Iraqi President Saddam Hussein to bend to the will of the UN Security Council and allow unhindered inspections by the UN Special Commission (UNSCOM). The crisis that blew up over this issue momentarily seized Congressional attention, but for the most part members of Congress freely pursued narrower political interests, which in many cases worked against the administration's internationalist objectives.

Partisanship Before Statesmanship

In the weeks leading up to the inauguration, a plethora of reports of possible illegal contributions to the President's campaign made it clear that scandals would again dominate Washington politics. The media focused first on the ethics violations of Speaker of the House Newt Gingrich, who had broken House rules by using tax exempt donations to his own election fund for party political purposes. Although the House Ethics Committee fined him $300,000, he succeeded in retaining his position as Speaker. A number of junior members of the House used the scandal and an embarrassing political defeat in June 1997 over a natural disaster relief bill to question his ability to lead. In July, Gingrich surmounted a plot to replace him, but the attempt damaged the Speaker politically and reinforced the impression that the Republican House was adrift without a rudder.

Gingrich's problems paled when set next to those of President Clinton, who faced three sets of allegations. Kenneth Starr, the independent counsel, continued the investigation of activities surrounding the Whitewater real-estate development in Arkansas. Then, in a highly publicised decision in June 1997, the Supreme Court ruled that the sexual harassment lawsuit brought against the President by Paula Jones, a former Arkansas state employee, could move ahead even while Clinton was in office. The more immediate set of allegations that the administration faced, however, were those over its fund-raising practices in the 1996 presidential campaign. These included accusations that the Chinese government had funnelled money to the President's campaign to counter Taiwan's superior influence in Washington, that the President offered perquisites, including stays in the Lincoln bed room of the White House and golf outings, in exchange for contributions, and that administration officials solicited funds on government time and from within government property.

In March 1997, the media revealed that Vice-President Al Gore had made fund-raising phone calls from the White House and had been dubbed 'solicitor-in-chief'. Gore's awkward effort, at a hastily called press conference, to limit the damage to his reputation as a politician of high integrity fell flat. It merely reinforced his image as a stiff and wooden personality. The whole episode increased the likelihood that other Democrats will challenge him for the next presidential nomination.

In January 1998, another scandal involving the President exploded in the media. Allegations that Clinton had a sexual relationship with a White House intern, Monica Lewinsky, who was 21 years old at the time, were vigorously denied by him, and also denied by Lewinsky in a sworn affidavit subpoenaed by the lawyers for Paula Jones. Lewinsky, however, had told a different tale to a friend, who had surreptitiously tape-recorded her reminiscences, and the persistent Starr seized this opportunity to add these allegations to the many others he has been investigating for over four years. The more serious charge against the President is that he lied about the affair and that he had suborned perjury by Lewinsky by advising her to lie about their relationship. By mid-March 1998, although Starr had interrogated many witnesses close to the President in the Grand Jury established to examine the charges, there was no indication that he would be able to make them stick. A vigorous campaign by the White House, Hillary Clinton, and supporters of the President claiming that Starr was motivated purely by political spite, and indeed was part of a right-wing conspiracy to bring down the President, seemed to be having considerable effect.

The stream of allegations failed to dent the President's popularity, which by mid-March 1998 had reached the highest level of his presidency. Many supporters perceived the fund-raising controversy as an attack by the President's political enemies. The Republican-led Senate hearings failed to convince them that there was much substance to the allegations of Chinese government efforts to buy influence or that the President or Vice-President had done much that the Republicans themselves were not also doing. Although many people felt that the President might well have had some relationship with Lewinsky, a large number were surprisingly forgiving, and more felt that Starr had gone too far in his aggressive campaign to prove the President guilty of suborning perjury.

While this belief partially explained Clinton's popularity and the public's willingness to overlook the allegations, the more likely reason was the booming economy. Indeed, in the first three months of 1997, the Gross Domestic Product (GDP) increased by an annualised 5.8%, a level of growth which has not been seen in the US since the 1960s. Figures for the last three quarters of the year indicate similar, albeit slightly lower, levels of growth. Unemployment dropped to about 4.7%, which several economists said was the natural level of unemployment in the country. At the same time, the stock markets climbed steadily during 1997, with the Dow Jones Industrial Average of blue-chip stocks up by more than 20%.

Such economic growth translated into vast increases in federal tax revenue, which in May 1997 created one of the conditions for the agreement on a balanced budget between the Republican Congressional leaders and the President. Political conditions for the agreement had been in place since the budget battle in the winter of 1995–96, when the President accepted the idea of a balanced budget and Congressional Republicans recognised that they had to avoid politically damaging government shutdowns. Both sides took the 1996 elections, which maintained a politically divided government, as a message that the American people wanted its branches to work together more cooperatively.

The agreement, finalised in August 1997 when Congress passed legislation, was supposed to lead to a balanced budget by 2002. Both sides could claim political victory for their role in the agreement. The Democrats stressed provisions that would allow tax cuts to defray the cost of college tuition and extend health care cover to children of single parents who lack cover. The Republicans emphasised reductions in taxes on capital gains and a $500 tax credit to families. Although the agreement passed through Congress with overwhelming margins, House Minority Leader Richard Gephardt voted against the final measure, in what many described as an attempt to distinguish himself from the Vice-President in hopes of challenging him for the Democratic presidential nomination in the year 2000.

The Good Students

The Clinton administration went through a learning process in foreign policy in its first term. Clinton and his advisers came to important decisions after tortuous, uncertain deliberations, and only gradually, in such areas as the Middle East and Bosnia, learned that it would only gain political points if it wielded US power judiciously. Buoyed by the strong economy and strengthened by the appointment of Madeleine Albright as Secretary of State, the administration demonstrated during the first year of the second term that it could put this important lesson into practice. On several fronts, including European and Asian policy, the administration confidently asserted its power. Congress, however, continued to limit the administration's freedom to conduct foreign policy, and the administration was reluctant to stand up too forcefully against what it considered to be public and Congressional opinion.

Nowhere did the growing confidence of the administration show itself more clearly than in Europe, where the US led NATO in the moves to enlargement, internal restructuring and intervention in Bosnia. At the beginning of 1997, the administration focused its attention on the largest obstacle to NATO enlargement, Russia. During her first trip as Secretary of State in February, Albright paved the way for a March summit between President Clinton and Russian President Boris Yeltsin by proposing an update to the Conventional Forces in Europe Treaty and the establishment of a joint NATO–Russia security council. At the summit itself in Helsinki, Clinton received assurances from Yeltsin that Russia would not oppose enlargement. The two also agreed to postpone the destruction of warheads under the second Strategic Arms Reduction Treaty (START) II for one year and to begin negotiations to reduce nuclear warheads from 2,500 to 2,000. At the July 1997 NATO summit in Madrid, the US won backing for its conception of the first round of NATO enlargement, overcoming widespread European support for the inclusion of Romania and Slovenia along with the eventual invitees, Poland, Hungary and the Czech Republic.

Bosnian policy again became the focus of internal bureaucratic battles in 1997. After announcing the establishment of the NATO-led Stabilisation Force (SFOR) in December 1996, Clinton promised that all US forces would leave by June 1998. Secretary of Defense William Cohen, a sceptic about Bosnian intervention since his tenure in the Senate, argued that Bosnia was mainly a European problem and repeatedly stressed that US forces would withdraw by the deadline. Two veterans of the tortuous deliberations that led to the 1995 intervention, Secretary Albright and the new National Security Advisor, Sandy Berger, opposed the withdrawal and gradually won Presidential support for their position. While it was unlikely that Congress would unilaterally revoke funds if the President decided to extend the mission, many of its members said Clinton needed to mount a public campaign to popularise the idea. In December 1997, Clinton took the first step, announcing his decision to maintain forces in Bosnia and conceding that the previous policy of setting a withdrawal deadline had been counterproductive.

Although the European allies welcomed US assertiveness on enlargement and Bosnia, they stridently opposed US laws penalising foreign firms that do business with the US's enemies. When the EU filed a formal complaint with the World Trade Organisation (WTO) about the Helms–Burton legislation on Cuba, State Department officials claimed that, because Cuba was a national security issue, the US would not cooperate with the WTO. The US and the EU compromised in April 1997 before the crisis could affect the approaching July NATO summit. Europe agreed to withdraw its complaint in exchange for a US agreement to postpone implementation of the legislation while common guidelines on foreign investment were negotiated. The August 1996 Iran–Libya Sanctions Act (ILSA), which imposed penalties on any company investing more than $40 billion (later $20bn) in Iran or Libya, was another source of inter-allied contention. In late September 1997, the French oil firm Total SA (together with Russia's *Gazprom* and Malaysia's *Petronas*), signed a $2bn deal with Iran to explore the South Pars natural gas fields. This deal seemed certain to trigger ILSA's sanctions regime, a step which the Europeans threatened would provoke them to relaunch their case with the WTO.

The Total deal was just one area of contention between Paris and Washington, and by the end of 1997 Franco–American relations had significantly deteriorated. France resented the US monopoly over the Middle East peace process and believed that Washington encroached on its sphere of influence in Africa when it supported Laurent Kabila in his rise to power in Zaire/Congo. Further indignation over the rejection of French proposals for enlarging NATO and the US refusal to relinquish command of NATO's military forces in the south resulted in the most telling symbol of the chilly relations, France's indefinite postponement of its plans to rejoin NATO's military command structure.

Washington's hard line on Iran was a key feature of US policy throughout 1997, but it began to be questioned towards the end of the year. After the election in August of Ayatollah Mohammad Khatami, an apparently moderate religious figure, as President of Iran, the administration said that it would take a wait-and-see approach. It was a short wait. In November, the US wrote a letter to Khatami that broached the idea of discussions on the countries' differences. In January 1998, Khatami gave an interview to Cable News Network (CNN) which suggested an interest in cultural exchanges as a first step toward improved relations. While noting that what was really needed was a change in Iranian foreign policies, the US welcomed this approach. The administration had clearly decided that it would be cautious, but open to any possible changes in Iranian attitudes.

An additional reason for trying to warm relations with Iran was the desire to isolate Iraq. In late October, seeing an opportunityfurther to erode the Gulf Coalition, Saddam Hussein expelled Americans on the UNSCOM inspecting teams and threatened to shoot down U-2 surveillance aircraft. The UN inspectors believed that they were on the verge of uncovering damaging evidence on the Iraqi biological weapons programme. Clinton dispatched a second aircraft carrier to the region and promised an overwhelming response to any attacks on US aircraft. Although Iraq backed down a week later, the diplomatic victory was tainted by the breathing-space for Iraq to remove evidence, the high-profile trip of Russian Foreign Minister Yevgeny Primakov, the explicit opposition of the Arabs to military action, and the fact that Saddam continued to tweak UNSCOM in the beginning of 1998.

Saddam Hussein wasted little time in testing the will of the Security Council once again. In January 1998, he refused to allow UNSCOM access to what he called 'presidential sites', but which the inspectors believed were factories and buildings where the Iraqi regime stored biological and chemical agents. The US, backed by the UK, increased the number of aircraft in the region and threatened heavy damage if the regime would not agree to abide by the outstanding Security Council resolutions requiring full access by inspectors to all suspect sites. At the last moment, after a trip to Baghdad by the Secretary-General of the UN Kofi Annan, an agreement was signed that temporarily lifted the threat of military action. The US and UK, however, maintained their forces in the Gulf to assure Saddam's compliance with the agreement (see map, p 227).

The rejuvenated diplomacy of the Clinton administration did little to change the moribund Middle East peace process. On the contrary, the US role as mediator slowly degraded. When the Israelis began construction of settlements in Arab east Jerusalem in March 1997, the US cast three lone vetoes in the Security Council against resolutions that condemned the construction. Although US Representative Bill Richardson argued that the Security Council was not the proper forum to discuss issues that would arise in the final-status talks, Palestinians saw the votes as overt support of Israel's interests. Such perceptions increased during the trips to the region by Special Envoy Dennis Ross, as Palestinians accused him of taking Israel's position in the peace talks. Despite clamours by Palestinian negotiators for higher-level involvement, Albright avoided a trip to the region until September, when she called for a 'time out' in settlement construction. She also asked Israeli Prime Minister Binyamin Netanyahu to abide by the provisions of the January 1997 Hebron Accord, which included two Israeli withdrawals in 1997. She asked the Palestinians to combat terrorism, calling this fight the essential link in the agreement. Her attempts to convince the leaders of countries in the region to stick to their plans to attend the Middle East–North Africa Economic Conference (MENA) in December in Qatar failed. Although Arafat and Netanyahu both visited Washington in December to talk with Clinton, little progress was made in getting talks underway again.

The Clinton administration had some success during 1997 in rebuilding its relationship with China, from the nadir reached in March 1996, on the basis of mutual interests in security, trade and curbing nuclear proliferation. The most promising results were achieved in the second area.

The US trade deficit with China was reduced and bruising battles in Congress over the renewal of China's Most Favored Nation trading status were avoided. During a March visit to Beijing, Vice-President Gore attended a ceremony completing sales of Boeing aircraft to China and establishing a joint venture with General Motors. Chinese President Jiang Zemin's October 1997 trip to Washington went smoothly, and included the conclusion of a deal with Boeing for an additional $3bn sale, consolidating Boeing's position as the leading aircraft provider to China. The Chinese agreed to end their support for Iran's nuclear weapons programme, which led Washington to offer China previously embargoed nuclear energy technology. And during a visit by Defense Secretary Cohen to Beijing in January 1998, he was allowed into a hitherto secret air defence base in Beijing, signed a naval pact aimed at avoiding possible conflict between US and Chinese warships and was given assurances that China would cease supplying anti-ship cruise missiles to Iran.

The two countries failed, on the other hand, to reach any agreement on problems relating to human rights or attitudes toward Taiwan or Tibet. There were still many areas which could create considerable trouble for both countries. But, by emphasising those areas where cooperation seemed possible, and putting to one side those areas that could lead to disagreement, the badly strained atmosphere that had threatened the relationship was lightened. The administration could count this as one of the foreign policy successes of 1997.

Constraints On Defence Resources

At the beginning of 1997, the Defense Department was faced with a new Secretary, former Republican Senator William Cohen, and a new review of defence requirements, the Quadrennial Defence Review (QDR). Cohen came into office with considerable experience in defence policy, having served on the Armed Services Committee, but little experience in administration or management. The Quadrennial Review, which had already begun under Secretary William Perry, was Cohen's first management task as Secretary.

During Clinton's first term, Secretary Les Aspin had supervised his own review, The Bottom-Up Review (BUR) and overseen the first Clinton budgets which reduced defence resources from the projections during the previous administration of George Bush. They had set out a difficult course for defence planning. The BUR had posited two major regional wars (The Gulf and Korea) as the framework for defence planning and resources. But the review did not go very deeply into the way these contingencies should reshape US forces, nor did it do much to tackle the large defence infrastructure inherited from the Cold War. Most seriously, the BUR framework left contingency operations to be dealt with using residual capabilities acquired as a result of building forces for the two big wars.

From 1993 on – in Somalia, Haiti, Rwanda and ultimately Bosnia – the armed services learned valuable lessons about the actual uses to be made of US forces, and these lessons centred on what was only residual in the BUR. These contingency operations cost money that was not budgeted for in the original defence plan, which had reduced defence budgets by nearly 40% in constant dollars from the 1985 peak in the Reagan administration. To cope with these costs, and to ensure high levels of readiness and troop morale, the Clinton administration adjusted its defence budget plans upward four times between 1993 and 1997.

By 1997 it was clearly necessary to rethink both defence requirements and the resource plan. Policy-makers had to cope with the likelihood that, in the future, forces would be used most frequently as the contingency operations undertaken in the recent past had indicated. They also had to deal with the fact that, since 1988, large, ready forces had been made possible only by draconian reductions in spending on defence procurement.

Procurement funds had fallen nearly 70% in constant dollars between 1985 and 1997. The budget basically funded ready forces at the cost of modernisation, with the consequence that more than half the inventory of tanks, helicopters and aircraft were more than half-way through their expected lifetime. In addition, the Gulf War and the subsequent research into 'total battlefield dominance' suggested that a new generation of defence technology was on its way and needed to be integrated into future procurement plans. All three services were beginning to implement this 'Revolution in Military Affairs' (RMA) – the Air Force with the B-2 bomber, cockpit communications and smart missiles, the Army with the decision to move toward a 'digital battlefield', and the Navy with a new generation Tomahawk missile carrying improved guidance. The Joint Chiefs, under the strong leadership of Vice-Chair Admiral William Owens, had also begun to move in this direction.

The Department had to cope with these changes, while recognising that defence resources were unlikely to grow significantly. In 1995 and 1996, the Republican Congress had reacted to Clinton defence budget proposals by increasing funding (by more than $11bn in 1995 and another $6bn in 1996), much of it for procurement. By 1997, however, both the White House and Congress were fixated on achieving a balanced budget. As the QDR proceeded, the Republicans in Congress seemed as anxious as their Democratic counterparts to continue sustained reductions in overall defence budgets. Ironically, it was the administration which insisted that, where defence was concerned, the budget deal should stick to the original Clinton plan, which stopped reducing funds in 1998 and, in particular, increased procurement spending by 40% in constant dollars from 1997 to 2002.

The budgetary negotiations and the requirements review left the Department somewhat better off by the end of the year, but the outcome did not satisfy everyone. The QDR explored three options for overall planning guidance: one that emphasised large forces in the near-term, at the cost of investment in technology; one that emphasised the RMA and paid for it by cutting forces; and a third that did a bit of both – maintaining slightly smaller forces while hedging with hardware and investment in technology. The third option carried the day.

The QDR concluded that the two-major-war scenario of the BUR should be maintained, since tensions did not appear to have relaxed sufficiently in either the Gulf or the Korean theatre to permit much smaller US forces. It did, however, add that dealing with smaller-scale contingencies should be elevated to a more significant planning requirement than had been the case in 1993. This added some to the defence burden, rather than simplifying the problem, since even two such small contingencies would put greater stress on a slightly smaller force.

The review also concluded that the target of $60bn for defence procurement was still correct, but could only be attained a year later (2001) than originally planned. Although the services would still be under great pressure to restructure their current modernisation plans, they would not have to cancel any major system. The Air Force could continue with its plan to procure the next-generation F-22 stealth fighter, though it would not be able to buy as many as it would have liked. The decision made it impossible to add any more B-2 bombers, but it allowed the service to continue to acquire the C-17 transport, which is critical to future airlift requirements for contingency operations. The Navy shuffled the pieces of its shipbuilding programme around so that it could continue to buy the next carrier, at least three guided missile destroyers a year and an amphibious vehicle. Navy leaders, however, felt that their shipbuilding plans were too tight and that the projected construction plan would not sustain more than a 300-ship Navy over the long term. The Army suffered the most from the squeeze, with an $8bn acquisition programme dedicated largely to tank upgrades, new artillery and the purchase of a large number of support vehicles.

It was not easy for the services to squeeze in hardware while still maintaining the required high degree of readiness and a large overall force (no divisions or battle groups were cut, while the Air Force moved some aircraft to reserve wings). The way out of this dilemma was, inevitably, the time-honoured one: institute reforms in defence management. Since 1993, under budgetary pressure, the Defense Department had done rather well in reforming some of its management structures: consolidating finance facilities, reducing more than 20% of the base infrastructure through four rounds of base closures, and reducing civilian employees by more than 200,000. The new stringency meant more of this to come, and roughly $3bn a year was found by consolidating support activities, reforming business practices (including electronic contracting and smart card buying), privatising such operations as personnel and payroll, energy supplies and logistics, and by reducing positions in the various defence agencies reporting directly to the Secretary of Defence.

Gaining these management and infrastructure savings will not be easy. In 1997, the Department asked for the authority to carry out two additional rounds of base closures. Although such closures do not produce immediate cash (they cost construction funds in the near-term), they are critical to long-term savings. Congress refused this request. It had been dismayed by the way Clinton had politicised the 1995 closure round (the administration insisted on keeping logistics centres in Sacramento and San Antonio open under private management) and was unwilling to pay the near-term political price of closing more bases. A similar request made in 1998 faced similar resistance.

The QDR has met with mixed reviews. It was roundly criticised by the independent National Defense Panel, set up by Congress, as too timid on force cutting, unrealistic on management savings and inadequately committed to the RMA. On the other hand, Congress itself has seemed more receptive. The administration's 1998 budget request was largely approved, with only a small addition of roughly $3bn (1%). The bill included $3.4bn for the construction of four AEGIS destroyers, and $2.3bn for the first of the next generation attack submarine. RMA advocates, concerned about the lack of long-range strike capability, were infuriated when the House–Senate conferees reduced funding for development of the automated arsenal ship, which the Navy cancelled in October, and allowed the administration to deny a $331m down payment for nine additional B-2 bombers once it reviewed the report of a special panel in early 1998. Yet, to ensure continued recruiting and support morale, the bill also increased basic pay by 2.8% and established a range of new retention bonuses to help keep Air Force pilots from leaving the service.

Reaching Out

If, like most Presidents, Clinton wishes to leave a lasting legacy, he will have to advance a number of significant initiatives in the final three years of his administration. Yet he faces considerable constraints in trying to do so. Above all, he must ride out the storm of allegations of sexual mis-demeanours and financial peccadilloes. Beyond that, he needs to tread a fine line, conciliating his fellow-Democrats in the House, while not ceding any political advantages to his Republican opponents. As a result, he has chosen to move ahead in graduated steps, avoiding over-arching programmes in favour of small advances on such issues as health care, child care and education. These worthy efforts will not go down in history as leaving the kind of mark left by Rooseveltian or Johnsonian initiatives, but they provide satisfaction and considerable reward for many ordinary citizens. In any event, since Clinton is not faced with any large challenges, and must deal with a divided government, they are the best he can hope for with regard to domestic concerns.

It is ironic that it is in the foreign arena that the President may have the greatest opportunity for a legacy. Yet his interest in this field is still sporadic and the focus of his attention is still clearly on internal developments. To create a legacy, the President would have to mount sustained efforts to overcome the narrower interests of Congress and involve the American people in order to create the domestic constituency for stronger initiatives. Such concentrated activity could lead to the settlement of many issues in the Middle East, Europe and Asia, but this would require both greater vision and consistency of purpose than the administration has yet shown. The likelihood is that the President will make minor progress on his domestic agenda after intense domestic diplomatic efforts, and manage crises abroad as and when necessary, but will have only the fact of NATO enlargement as a lasting strategic accomplishment.

Latin America: Balancing Military And Civil Powers

It became evident during 1997 that, while the generals in the region's democracies have returned to their barracks, not all have given up their designs on power. This was underscored with the election of General Hugo Banzer, Bolivia's former iron-fisted dictator, to the presidency in May 1997. Banzer's return to power through the ballot box was not an aberration; throughout the continent former military officers were taking advantage of the instruments of democratic governance to participate in the political life of their countries.

In Paraguay, incarcerated General Lino Oviedo has indicated that he will stand in the May 1998 presidential elections. In Colombia, former General Harold Bedoya, ousted by Ernesto Samper's government earlier in the year because of his hardline opposition to negotiations with the guerrillas, has emerged as an independent presidential candidate and has climbed as high as second in public opinion polls. Although unlikely to win the May 1998 presidential elections, Bedoya has galvanised a right-wing constituency which feels under-represented in the two principal parties, the Liberal Party and the Social Conservative Party. In Venezuela, former Lieutenant-Colonel Hugo Chavez, whose support dates back to a popular but failed coup attempt in 1992 against the government of disgraced President Carlos Andres Perez, has also emerged as a leading candidate for the November 1998 presidential elections in his country.

Former and current military officers are not only emerging as presidential candidates, but they are wielding considerable influence behind the scenes of democratically-elected civilian governments. Some of this is a result of earlier constitutional and political manoeuvres by outgoing military governments intent on maintaining their power. The continuation of General Augusto Pinochet as Commander-in-Chief of the Chilean Armed Forces from 1990–98, with control over military budgets and personnel decisions, is an example of this. Under the 1980 constitution, written by General Pinochet and still in force, the former dictator was allowed to assume the position of Senator-for-life in the Chilean upper house the day his term as military chief ended in March 1998.

Military officers are also assuming power in less institutional ways. In Mexico, the armed forces are increasing their influence as they take on a wider role in the fight against the Zapatista guerrillas, and against the five major drug cartels that have grown up in the country. In Colombia, the military have long wielded great power at the regional and local level, notably in the principal zones of violence dominated by guerrillas, drug-traffickers and the newly-powerful paramilitary squads. In Ecuador the armed forces remain one of the most popular institutions in the country – a legacy of their role as modernisers during the heyday of oil revenues in the 1970s. The military proved to be the major arbiter of power during the showdown between President Abdalá Bucaram and Congress in February 1997, which led to his being ousted. In Peru, President Alberto Fujimori has come to rely excessively on the armed forces and the state intelligence agencies both to govern and to concentrate power. Following the successful rescue operation on 22 April 1997 that freed Japanese, Peruvian and Bolivian hostages held by Peruvian guerrillas in the Japanese Ambassador's residence, the prestige and power of the armed forces grew. Within a few months, as Fujimori's popularity began to wane steadily, the President came to rely even more on this bastion of institutional support.

Events in 1997 underscore that civil-military relations within a democratic Latin America have taken a new turn: the threat to civilian rule no longer comes from coup-making militaries. Rather it derives from current and former military men who are able and willing to step into the institutional, social and political power vacuums and use democratic instruments and procedures to further their authority. This is occurring even as the region consolidates the practices of regular elections, and

continues to achieve steady rates of economic growth. Yet the same developments are also aided by escalating crime rates throughout the region; the continued ferment of guerrilla insurgency in Colombia, Peru and Mexico; the explosive appearance of paramilitary squads in Mexico and Colombia; and the spreading tentacles of organised crime – most notably that related to the supra-national but locally-rooted drug trade. The armed forces may be back in the barracks formally, but their power and influence still resonates throughout the region's democracies.

Struggle For Democracy And Sovereignty In Mexico

As much as any country, Mexico reflects the dual tendencies towards greater democracy and greater militarisation of politics which are both due to the growing security challenges facing post-Cold War Latin America. The monopoly of the ruling Institutional Revolutionary Party (PRI) was effectively broken, after almost 80 years, by the elections in July 1997. The PRI won only 38.4% of the vote. The two leading opposition parties, centre-left *Partido Revolucionario Democrático* (PRD) and centre-right *Partido de Acción Nacional* (PAN) gained 25.8% and 26.9% respectively. Two other parties won an additional 15 seats, bringing the combined opposition vote to 261 in the 500 seat Federal Chamber of Deputies. Moreover, PRD leader, Cuatemoc Cárdenas won the first-ever mayoral election for Mexico City. These results pave the way for a major challenge to the PRI's hold on the presidency when those elections are held in 2000.

Immediately following the polls it became clear that politics in Mexico were no longer the same. As the sessions in the legislature advanced, it became clear that holding together an opposition coalition would be difficult. This emerged when wrangling broke out over the 1998 budget. Centre-right PAN deputies ultimately broke ranks with the PRD and allied themselves with the ruling PRI. The political situation is one of shifting power alliances among three strategic actors. The PAN and PRI are closer ideologically, but PAN and PRD see an advantage in allying to wrest institutional power from the PRI.

Even as Mexico welcomed the first experience of multiparty democracy, the country was being battered by the pernicious effects of the rise of drug trafficking cartels, and the expansion of guerrilla activity in Chiapas, Oaxaca, Guerrero and other states. The changing security picture has brought the Mexican armed forces into the political arena to a degree that had no precedent during the long era of PRI one-party rule. Senior army officials have been linked to drug trafficking as have scores of lower-ranking commanders and police chiefs. At the same time, the Mexican Army has dramatically expanded its spending on weapons and recruits. Mexico now has its second-largest number of serving military personnel ever: 175,000 in the army and navy and 8,000 in the air force. Over the past three years, national defence spending has increased by 1,000% to $1.58 billion, reflecting more the military's previous institutional weakness than its current political strength.

Much of this spending has been to fight narcotics and to contain guerrilla insurgency. Yet increased military activity has led to a dramatic upsurge in human rights violations. Paramilitary units have engineered several massacres of Indians and peasants believed to support the Zapatistas in insurgent zones such as Chiapas. Moreover, the US General Accounting Office, an agency which monitors government spending, has charged that equipment given to fight drugs, such as helicopters, armoured personnel carriers and night-vision equipment, is being used principally against leftist insurgents. There is concern that US anti-narcotics aid will contribute to a rise in human rights abuses.

The Andean Region: War, Drugs, Weak Democracies

The nations of the Andean region continued to struggle with the insidious effects of the drug trade on their societies, economies and political systems; coping with US anti-narcotics policies; and dealing with the demands of troubled democratic politics. In all of the countries, democratic institutions have been faced with severe challenges. In Peru, 1997 began with the country in the full throes of an international crisis. On 17 December 1996, the *Movimiento Revolucionario Tupac Amaru* (MRTA) staged a spectacular assault on the Japanese Ambassador's residence, seizing over 700 hostages, including over a score of ambassadors. For several months, President Fujimori acted with great prudence, although he repeatedly declared that he would not give in to MRTA's demands for the release of all political prisoners and safe conduct back to their bases. On 22 April 1997, almost four months after the crisis began, Peruvian commandos stormed the residence from underground tunnels, rescued all the hostages and killed all the guerrillas. Fujimori's popularity, which had fallen below 40% in the opinion polls in the weeks following the hostage-taking, rebounded and rose to 74%.

His popularity was not to last. In January, three members of the seven person Constitutional Court had ruled that Fujimori could stand for re-election for a third term, but four members had abstained on the issue. The head of the Court publicly declared that the abstentions meant that Fujimori was ineligible to run, and the government-controlled Congress then voted to remove three of the judges who had abstained. The Peruvian public had been tolerant of Fujimori's authoritarian tendencies when these were used to fight terrorism or restructure the economy. They proved less tolerant when power was abused for the President's personal gain. By May, Fujimori's standing in the polls had precipitously fallen to just 25%. Yet he continued to enjoy the support of his senior military and intelligence advisors, and it was clear that he felt this was of greater importance.

Fujimori also found another major ally in the US by becoming a willing participant in Washington's war on drugs. Over the past few years, sophisticated radars have been placed in Colombia and Peru to track aircraft carrying narcotics. Fujimori has given the Peruvian Air Force permission to shoot down suspect planes. The US and Peruvian policies have effectively shut down the air-bridge between Peru and Colombia, leading the US to give public praise to the resulting 18% drop in Peruvian coca production as prices fell.

Colombia: Not The Gem Of The Ocean

Yet past experience suggests that if drug production falls or is eliminated in one place, it will soon emerge in another. The closure of the Peru–Colombia air-bridge has simply pushed coca production north into Colombia. There, despite a US government assisted programme that sprayed 16,000 hectares of coca in 1996–97, overall coca cultivation actually increased by 32%. Colombia is now not only the world's largest cocaine producer, it is also the largest coca producer, representing a major restructuring of the Andean narcotics business. Further, Colombia has become a significant producer of heroin and now supplies up to 60% of the US market.

This has all happened while US–Colombian relations have been severely strained by Washington's certification programme, under which the US President must certify that states receiving US aid are cooperating in the anti-narcotics war. In both 1996 and 1997, Colombia was 'de-certified', while in 1998 it was de-certified but given a national security waiver. In the Colombian case, de-certification is mainly symbolic, expressing a vote of no-confidence in President Samper whom the US accuses of taking money from the Cali Cartel to finance his 1994 election campaign. Yet, even as official relations soured, the US continued to work with the Colombian military, police and judiciary. Anti-narcotics aid actually increased during this period, reaching over $100m in 1997. Most of this is earmarked for the military and judicial systems.

Because of concerns about human rights, US anti-narcotics aid channelled away from the army in 1994 and directed to the police and judiciary. The US had evidence that units receiving the aid had been responsible for some of the worst abuses of human rights in recent years. Military aid was resumed in 1997, however, and the new US policy has pushed the army back into the anti-narcotics business.

The other notable feature of 1997 in Colombia was an escalation of violence. The guerrillas have launched several major attacks , one on a lone military base in the southern jungles where they captured 60 soldiers. In March 1998, *the Fuerzas Armadas Revolucionarias de Colombia* (FARC) launched one the biggest offensives of the 30-year war, decimating one of the Army's élite counter-insurgency units. The general impression is that the military is losing the war and that the guerrillas have gained new offensive capabilities, although there are still major doubts whether they could achieve a military victory.

As the military has grown weaker, however, the loosely-controlled paramilitary forces have grown stronger. In October, one of the major paramilitary leaders, Carlos Castaño, announced the formation of the *Autodefensas Unidas de Colombia* (AUC). The paramilitaries mainly target the civilian population, seeking to undermine the guerrillas' social base. Many of them operate in close coordination with the army and police. Others are more independent. However, in the absence of an effective state presence, the paramilitaries have begun increasing the range of their activities. In July, the AUC sent forces half way across the country to the town of Mapiripán, in the eastern plains, a guerrilla stronghold, and massacred over 30 people in an operation that lasted for eight days. Mapiripán was turned into a ghost town, and more than 1,000 refugees joined the growing army of internally displaced people in Colombia. There are now over one million internal refugees in Colombia.

The spreading violence in Colombia has attracted the attention of the international community. The United Nations, the World Bank and several governments, including Spain and Costa Rica, have begun to look for ways to help Colombia confront the escalating conflict. The emerging consensus is that the violence is much more than an anti-narcotics issue. Drug revenues are fuelling it, and help support both guerrillas and paramilitaries. However, there are deep social and political fissures in Colombia as well as state and institutional weaknesses which must be addressed, as similar problems were tackled by the peace processes in Central America.

Venezuela and Ecuador

Venezuela has been particularly affected by the fighting in Colombia. One of Colombia's guerrilla movements, the *Ejército de Liberación Nacional* (ELN) which operates in areas along the border between the two countries, has continually crossed into Venezuelan territory and engaged Venezuelan National Guard in combat. In response, Venezuela purchased six surveillance planes from Poland to control its porous border. The government has also requested help from the US in a move which greatly disturbs the Colombian authorities. Venezuela has now stationed 30,000 troops on its border with Colombia.

Venezuela continues to suffer politically from institutional weaknesses and the unpopularity of economic adjustment measures enacted by the government of Rafael Caldera. Although the government has stated that the economic stabilisation programme, known as Agenda Venezuela, is beginning to show signs of success, most of the population seems to disagree. On 6 August 1997, millions of workers heeded a call for a one-day general strike, which brought the country to a standstill, without, however, resorting to violence.

Having dipped into recession in 1996, the economy rebounded in 1997, growing at over 4%. However, inflation still remains high at 38.2% and officially-reported unemployment is 12.4%, with underemployment much higher. Poverty levels continue to climb, crime has increased sharply and corruption continues without restraint. Despite these circumstances, Venezuela's traditional parties have not been able to make a comeback. Leading in the opinion polls for the presidential balloting scheduled for

December 1998 is former Miss Universe Irene Sáez, currently the Mayor of Chacao. Ten points behind her is Lieutenant Colonel (rtd) Hugo Chavez who was imprisoned for his coup attempt in 1992 but pardoned by President Caldera. He is again trying to create a movement which unites those displaced and impoverished as a result of Venezuela's economic crisis.

Ecuador has been able to avoid the violence and narcotics-influenced politics of its neighbours, but has not avoided the institutional instability. In February 1997, its unicameral Congress unseated elected President Abadalá Bucaram, claiming he was 'mentally incompetent'. At first, both the Vice-President Rosalía Arteaga and the President of Congress Fabio Alarcón laid claims to the presidency. Arteaga was sworn in for two days, but, when the armed forces supported Alarcón, he was named interim president until August 1998.

On 25 May 1997, two months after the congressional coup, Alarcón submitted a plebiscite to the people, asking whether the electorate supported the removal of Bucaram and whether they approved the designation of Alarcón as interim president. The first question was answered affirmatively by 74% of the voters and the second by 65%, in effect legitimising Alarcón's presidency until August 1998. The plebiscite also endorsed the convening of a Constituent Assembly to rewrite the constitution, and elections for this body were held in November. The voting results did not substantially alter the correlation of forces in Ecuador, nor does anyone expect the revised constitution fully to address the institutional crisis.

The nation was thrown into further institutional disarray when the Congress voted to remove the President of the Supreme Court of Justice, Carlos Solórzano and 30 other judges from the High Court. Congressional leaders claimed that they wanted to de-politicise the Court. Solórzano denounced the move as one more example of the Congress over-stepping its authority and announced that he would run for president in the presidential elections scheduled to be held in May 1998. He has little chance of winning, however. Jaime Nebot from the centre-right Social Conservative Party, who unexpectedly lost to Bucaram in 1996, leads the field, with independent Freddy Ehlers as his main challenger.

Nineteen years after leaving power, the military still remains extremely popular with the public and extremely powerful within the government. The armed forces regularly lead the list of institutions that the public supports, far out-polling the presidency, the Congress or the judiciary. In the weakly-institutionalised landscape of Ecuadorian democracy, almost any outcome is possible in the 1998 presidential elections, as the history of the last nineteen years has demonstrated.

Bolivia

In Bolivia, the former military dictator, Hugo Banzer who had seized power in a coup in August 1971 and ruled until 1978, when he was overthrown by a former loyalist in another coup, achieved power through the ballot box in June 1997. Banzer supports his predecessors' neo-liberal reforms but is less enthusiastic about some specific measures relating to de-centralisation, indigenous rights and the earmarking of funds from selling off state industries for a national pension fund. All these reforms were trademarks of the previous President Gonzalo Sánchez de Losada, and the question remains whether Banzer will reverse or support them.

Internationally, Banzer vowed that he would rid the country of illegal coca production during his five-year term and would cooperate with the US. He declared that his government would annually eradicate 7,000 hectares of coca plants through voluntary programmes offering each farmer US$2,500 per hectare if they abandon coca cultivation and through expanded development assistance. The US earmarked $114m for anti-narcotics and alternative development in Bolivia in 1997. In the past, the coca-growing zones have been the major source of conflict and human rights abuses. In response, the coca growers have become highly organised and have been able to gain political power at the local level. In the June elections, the leader of the largest federation of coca farmers, Evo Morales, was elected to Congress. If La Paz's eradication programme is to be successful, it will require the cooperation of the farmers. The leaders of the coca growers say they are willing to work with the government as long as it follows up on its commitments to fund alternative development opportunities.

In general, with the closing of the air route to Colombia and increased government pressure, total coca production has been steadily declining. But even as Bolivian production diminishes, new crops are being grown over the border in Brazil and Paraguay.

Southern Cone And Brazil: Military Versus Civilian Power

In two nations of the southern cone, Chile and Paraguay, civilians and military authorities are engaged in a constant battle over power, even as other areas of national democratic life seem to be flourishing. In Chile, two successive governments have been elected under the rules devised by former military dictator Augusto Pinochet, who is still active at 82 years old. After handing over power in 1990, Pinochet stayed on as Commander-in-Chief of the armed forces until 10 March 1998 when he became senator-for-life. The country is divided over the continued role of the former dictator who ruled ruthlessly for 17 years. His government had been marked by widespread repression, killings, disappearances, torture and exile. Yet it was also credited with laying the foundation for strong economic growth and the most stable and modernised economy in Latin America.

The day of Pinochet's transformation from Army Commander to Senator was marked by street protests led by university students in downtown Santiago and by protests within Congress. Yet many, including President Eduardo Frei Ruíz-Tagle, feel that this was the necessary price of ensuring the transition to democracy that began when Pinochet handed over power after losing a plebiscite in 1988. For others, however, Pinochet's presence in the Senate, together with a block of 15 other senators appointed for life or specially designated by the military (out of a total of 38 senate seats), is a daily reminder that the civilian government is not able to stand up to the armed forces. The government is unable to secure enough votes to modify the constitution or re-open human rights cases from the Pinochet years.

President Frei has expressed his desire to eliminate what he calls the 'authoritarian enclaves' (the lifetime and designated senators plus the institutional autonomy of the armed forces) within the Chilean state. The Senate, however, has rejected reform measures on three occasions. With Pinochet now leading the military block in Congress, it is unlikely that reforms will be made soon.

From a strategic perspective, it was feared that Chile's military autonomy was leading to an imbalance of military power in the southern cone. Over the past decade, the Chilean armed forces have bucked regional trends and have substantially increased spending on the military. The IISS and other independent analysts have placed military spending at above $2bn. However, in a sign that civilian authority over the military is slowly beginning to emerge, President Frei and President Carlos Menem of Argentina signed a defence cooperation agreement in July 1997 and scheduled the first-ever joint military exercises between the two states. Although at the time Pinochet publicly questioned the agreement, he ultimately gave his support. Indeed the cooperative defence pact helped spur the US, on 31 July 1997, to lift its ban on sales of advanced weapons to Latin America.

Chile is at a cross-roads. It must modernise and democratise its political system, while at the same time maintain its economic dynamism. Although it is 25 years since Pinochet first took power, and eight years since he handed power back to the civilians, the post-Pinochet era has barely begun.

When Paraguay's top military officer, General Lino César Oviedo Silva, staged an abortive coup in April 1996, most viewed its failure as a triumph for democracy. Representatives from the Organisation of American States, the US State Department and the neighbouring countries of the Mercosur (*El Mercado Común del Sur*) trading community all rushed to Asunción to try to stem the crisis and assist President Juan Carlos Wasmosy to preserve constitutional rule. Oviedo was arrested then freed from preventive detention on 8 August 1997 pending trial for the coup attempt.

Events took a disturbing turn when Oviedo, with the charges of rebellion still hanging over him, announced his candidacy for president as the nominee of Paraguay's ruling *Asociación Nacional Republicana (ANR)* party. ANR has held power for over 50 years, largely as a result of its close relations with the armed forces. It was the party of long-established dictator General Alfredo Stroessner; it is the party of President Wasmosy.

Following a heated and closely-contested primary race in September 1997, the elections tribunal declared that Oviedo had won the party's nomination. The victory caused an uproar. President Wasmosy and others declared they would appeal the vote count; the chief judge of the Supreme Court announced it would continue the proceedings against Oviedo even if he became the official nominee. The US Ambassador publicly deplored the spectacle, lamenting that a man who had threatened the constitutional order 17 months earlier could become the ruling party's official nominee. When the ANR leaders refused to register Oviedo as candidate, he took the party to court to force them to respect the primary results.

Not waiting for the court to act, Oviedo immediately began taunting the President and his rivals, acting more like a president-elect than a candidate. Following certain inflammatory remarks, Wasmosy clumsily ordered Oviedo's 'disciplinary arrest'. Still the fiery ex-general vowed to continue his campaign through the elections scheduled for 10 May 1998. The showdown with the Wasmosy government seemed only to burnish his image as a populist who stands on the side of the people. In March 1998, the Supreme Court found Oviedo guilty of rebellion, throwing the presidential race into further disarray.

Whoever becomes the official ANR candidate will face long-standing opposition leader Domingo Laino of the *Partido Liberal Radical Auténtico*. It is the 64-year-old Laino's third attempt at the presidency. But Oviedo cannot be ruled out. As elsewhere in Latin America, in the new era of electoral politics, previous experience in coup-making has proved to be an effective way to enter the political arena.

A Better Picture in Brazil and Argentina

Democratic politics and civilian control over the armed forces is much more consolidated in Brazil and Argentina. Both countries, and particularly Argentina, have reduced military expenditure. Argentina has also been in the forefront of re-training its army for international peace keeping missions. Moreover, President Menem has signed defence cooperation agreements with both Chile and Brazil. With Brazil, the Argentine Defence Ministry has proposed a 'system of common security' which would allow joint action on terrorism, drug trafficking and other contemporary threats.

Actions in Argentina's 'dirty war' between 1976 and 1983, when an estimated 30,000 people were killed or disappeared, continues to haunt the Argentine military and the country. Despite an amnesty granted by Menem, domestic human rights groups and foreign countries whose nationals disappeared during that period, continue to protest and call for those responsible to be brought to trial. It is unlikely that the amnesty laws will be overturned, and, as the country slowly comes to grips with this tragic period, it seems to be building new civil–military relations with military operations which are more appropriate for the post-Cold War world.

While cutting defence spending, President Menem has also positioned Argentina as a key ally of the US in international missions. Argentina sent ships to the Persian Gulf in 1990, and has troops with UN peacekeeping operations around the world. As a reward, the US has named Argentina a 'strategic partner', joining Japan, South Korea, Australia, New Zealand, Israel and Jordan which have a similar relationship with the US.

Both politically and economically, Argentina is beginning to feel the effects of almost a decade of Menemism, rapid privatisation and exposure of the economy to international markets. Following the onset of the Asian financial crisis, the Argentine stock market plunged more than 25%. Income inequality has grown and many who feel left out of the new economy are beginning to organise. In local and regional elections held in October 1997, the country's two leading opposition parties, the *Unión Cívica Radical* (UCR) and the *Frente País Solidario* (Frepaso) forged an alliance which unexpectedly defeated Menem's *Partido Justicialista* (PJ). The PJ lost in the federal capital of Buenos Aires, as well as in important cities and provinces around the country. Although Menem still holds a Congressional majority, this may mark the beginning of the end of Menemism.

Brazil, too, was forced to confront the consequences of its greater exposure to the global economy when the Asian crisis erupted in the final months of 1997. Brazil was the most vulnerable Latin American country because of its overvalued currency, its trade deficits and capital inflows. President Fernando Henrique Cardoso had staked his political reputation on maintaining the value of the Brazilian currency, the *real*. The Asian crisis threatened to undo his work. On 12 November 1997 alone, the Brazilian stock exchange fell 10%. Cardoso acted quickly and adroitly. He immediately pushed through Congress a severe austerity package cutting over 33,000 public sector jobs, raising income taxes and doubling interest rates. He took $8bn from the nation's foreign reserves to defend the *real* on the foreign exchange markets. The austerity measures reassured investors and the Brazilian stock market rebounded, but the moves were not popular with the public and Cardoso's support in the polls fell to 27%, the lowest level of his presidency.

Despite the unpopularity of the austerity packages, the opposition seems unable to capitalise on Cardoso's difficulties. During 1997, he successfully presented a constitutional amendment to Congress permitting presidents to be re-elected for a further term, following similar constitutional changes in Peru and Argentina. By January, Cardoso's opinion poll rating had recovered to 39% and by the end of March he was again the front runner for the election in October 1998. Yet, because his popularity is closely associated with the country's economic stability – particularly his success in controlling inflation and the exchange rate – he remains vulnerable to unexpected forces from the international economy that are beyond his control.

Cardoso's weakest spot remains Brazil's severe social problems, particularly the upsurge in crime. Brazil has the highest rate of gun-related killings in the world, 26.9 for every 100,000 people (compared with the world average of 4.9). This translates into about 43,000 killings each year. In the 15–19 age group, the figure rises to 88.38 per 100,000. Murders increased by 18.5% in 1997. Cardoso has responded by increasing the police presence in the most violent shanty-towns, many of which are also major centres of both drug trafficking and drug consumption.

Critics and close observers argue, however, that the government's strategy has only aggravated the problem. Human Rights Watch/America has reported that police brutality and human rights abuses have substantially increased. Since the government inaugurated the new policies, civilian deaths at police hands have doubled to an average of 32 a month. This figure makes Rio's police among the most abusive in the world. Worse – and not unrelated – there has been an ominous growth in 'social cleansing squads'. Armed bands, which have been linked to the police, murder homeless people, beggars, street urchins and other 'undesirables'. These practices are now spreading well beyond Rio de Janeiro, with other cities, particularly São Paulo, now facing a dramatic increase in violent crime. This has led to a growing sense of insecurity in Brazil and to rising demands for effective government action.

Central America And The Caribbean: Post–War Societies

The end of the region's civil wars has meant a reduction in the armed forces and re-training of civilian police. Some former military officers, such as ex-Guatemalan dictator, Efraín Ríos Montt, hope to return to national power through the ballot box, but the general trend throughout the region is to curtail the influence of the armed forces and limit their institutional prerogatives.

Guatemala has embarked on a programme to reduce its forces by 33% since the country's last civil war ended in December 1996. In July 1997, President Alvaro Arzú surprised everyone when he sacked the two most senior army generals, the Minister of Defence and the Chief of the Defence Staff. The former had been a principal negotiator with the guerrillas; the latter was one of the principal opponents of the peace accord. The episode reflects Arzú's attempts to maintain a balance among competing army factions.

El Salvador and Nicaragua have also substantially reduced their armed forces since their civil wars ended in 1990 and 1991. When Guatemala completes its reductions, the region's militaries will be roughly in equilibrium. Panama and Costa Rica remain states without armies. Haiti also dismantled its army following the US intervention in 1994.

The armed forces of El Salvador, Guatemala, Honduras and Nicaragua have set up a coordinating body, called the *Conferencia de Fuerzas Armadas de Centroamerica*, in the hope of checking their declining influence in the region. The aim of the organisation is to redefine the role of the military in the post-cold war and post-civil conflict world. It is also seen by some as a partial advance toward the elusive goal of political integration in the region. One interpretation, however, is that the endorsement of the new military conference reflects the continued weakness of civilian authority in the face of military initiatives. Yet another interpretation is equally valid: civilian leaders recognise that it is better for the military to concern themselves with regional affairs than be tempted to meddle in domestic politics.

There is also an emerging consensus that each nation dismantled its police capabilities too rapidly following the negotiated peace accords. Crime has soared in each of the post-conflict societies, reaching alarming rates in Nicaragua, Guatemala and El Salvador. The latter nation currently has the second highest crime rate in the world, lower only than in South Africa and higher even than in Colombia.

At the same time, political violence in El Salvador has almost disappeared. In the new political environment, the former Marxist guerrillas, the *Frente Faribundi Martí para la Liberación Nacional* (FMLN) have successfully made the transformation from armed opposition movement to political party. In elections held in March 1997, the FMLN won 27 of a total 84 seats in the National Assembly. The ruling ARENA (Nationalist Republican Alliance) party won 28 seats. These are now the two largest parties in the assembly, transforming the principal military opponents of the 1980s into the two dominant congressional blocks of the 1990s.

Cuba has a long tradition of party control over its armed forces. This has not changed. However, since the country's economic crisis erupted in the early 1990s, the military has become more involved in domestic activities, ranging from from civil engineering projects to working in agriculture, and this has increased its institutional power. It is also reported that Defence Minister Raúl Castro, President Fidel Castro's brother, is assuming a greater share of the day-to-day burdens of government. During the congress of Cuba's Communist Party in October 1997, Castro specifically designated his brother as his successor. It is also notable that only one military officer was included in the 150 person politburo and the 24 member central committee. The new hierarchy is younger and contains more technocrats and economists, again reflecting Cuba's needs in a post-Soviet world. Cuba seems to be following the Chinese model: economic liberalisation first, democracy later.

Still Fragile, But Still There

Events in 1997 reflected the fact that democracy, although well-advanced, is still not fully institutionalised in the region. There is great concern about growing social inequality and the destabilising effects of economic and financial globalisation. Although the unsettled times are not triggering military coups, they do seem to be fostering a certain apathy toward the democratic process and some nostalgia for the old authority figures that the military once provided.

In this fragile democratic environment, former military officers and rulers are finding both their voices and votes. Their message is generally free-market and vaguely populist. It appeals to the fears aroused by the renewed cycles of insurgency in some countries, and the breathtaking rise of organised crime, corruption and insecurity throughout the region. Banzer in Bolivia, Pinochet in Chile, presidential candidates Bedoya of Colombia, Chavez of Venezuela, and Oviedo of Paraguay all represent a new strain of democratic politics throughout the region. It is one strain, but at the same time it is countered by the relative strength of political parties, social movements, other independent candidates and the increasing involvement of the international community. Democracy remains unconsolidated in the region, but it is, in most countries, still the most viable option.

Europe

West Europe had a good year in 1997. By mid-March 1998 it was not only clear that the European Economic and Monetary Union would come into being in January 1999, it also appeared that an unexpectedly large number of countries would join from the outset. This was the result of solid economic growth, but not enough of it to lower the average unemployment rate which still hovered around 12%. Europe did not have much success in constructing a common foreign and security policy, but it did move a long way towards expanding its institutions. NATO invited three new members into its club, and the European Union began negotiations with six candidate members. Turkey's EU application, however, was put on indefinite hold. This blow to Turkish pride, and the way in which the country's Islamic government was toppled, left most sections of Turkish society feeling pessimistic about the future and disillusioned with their political parties and political processes.

If the EU stumbled over Turkey, the UK became less of an obstruction when Tony Blair's Labour Party trounced the Conservatives in the May 1997 elections. A new cooperation replaced the old confrontation. Blair also reinstated close communications with the Irish government over the problems of Northern Ireland and reinvigorated the long-stalled talks on this issue. Despite a new surge of violence, mainly instigated by intransigent splinter groups on both sides of the fence, there was more than a glimmer of hope that an agreement acceptable to both Protestants and Catholics could soon be reached.

While affairs in most of Europe seemed to be in reasonable order, the same could not be said for Russia. There, in March 1998, President Boris Yeltsin demonstrated once again his penchant for erratic behaviour, sacking Prime Minister Chernomyrdin and his entire government and appointing a relatively unknown 35-year-old, Sergei Kiriyenko, as Acting Prime Minister. Yeltsin claimed that he was doing it because the old government had not been sufficiently energetic in pursuing economic reforms. In fact, the government had not been doing too badly. To be sure, it had not been able to pay all its workers on time, halt the slide in living conditions for many of its citizens, nor conquer the mounting wave of increasingly violent crime. Yet it had continued the vital privatisation programme, got a grip on inflation and begun some much needed reforms of the tax system. Yeltsin's move is sure to create instability in Russia in the short-run. But if it leads to greater verve in the government, this may be a price worth paying.

There was the threat of yet more turmoil in Yugoslavia in March 1998, as Serbian police and army moved forcibly against the majority Albanian population in Kosovo. The situation was critical and efforts by the US to restrain President Milosevic by increasing sanctions against Belgrade were being forestalled by French and Russian objections. In neighbouring Bosnia, however, NATO forces were effectively carrying out the Dayton Accords. A decision by President Clinton to leave US troops in the country for the foreseeable future was a further encouraging factor in the effort to secure peace, if not prosperity, there.

◆

Western Europe Moves Closer To Integration

Europe was startled in 1997 by two extraordinary election results. The Labour Party in the UK swept past the Conservatives in May, gaining power with one of the largest parliamentary majorities since the Second World War; while in France, in June, the Socialist Party under Lionel Jospin disrupted President Jacques Chirac's ill-conceived plans to use an early election to return to office with a stronger mandate and took power from the right. An election scheduled for September 1998 in Germany may well bring about a similar shake-up. European Economic and Monetary Union (EMU) not only stayed on track for its launch in January 1999, but an unexpectedly large number of countries seemed poised to make the target date. And the European Union (EU) summit in Amsterdam in June, which was called to make treaty reforms, while steering a minimalist course, cleared the way for negotiations on EU enlargement.

The steps taken to enlarge NATO, and its continued role in implementing the General Framework Agreement for Peace in Bosnia and Herzegovina (the Dayton Accords), affirmed the Atlantic Alliance as Europe's predominant security pillar and underlined the continuing role played by the US as a power in Europe. The EU, on the other hand, found it as difficult as ever to construct a Common Foreign and Security Policy (its 'second pillar'). Nor did the institutional revisions of the Amsterdam Treaty seem likely to change this. While European states did send 7,000 troops to Albania in April 1997, these forces deployed under national flags, not a European one. Europe was unable to speak with a single voice during the Iraq crises of late 1997 and early 1998, nor did any of the institutional innovation in the EU's second pillar help forge a common European policy toward the crises in Albania, Algeria or Kosovo.

Europe may have had difficulty acting as a unit in 1997, but its institutions were expanding. NATO's Madrid summit in July and the EU's Luxembourg summit in December set the two organisations on the path to enlargement. Moreover, 11 Union members were set to embark on EMU. Solid economic growth in the EU in 1997 (2.6% against 1.4% in 1996) helped to bring fiscal deficits below Maastricht's 3% ceiling – creative book-keeping did the rest. While this growth meant more countries would qualify to join EMU in the first round, it did not significantly lower the Union's unemployment rate, which, in fact, rose and was still averaging above 12% in March 1998.

Germany Struggles With Reform

Germany's longest serving chancellor, winner of four elections and in power for 16 years, Helmut Kohl found himself buffeted by political troubles in 1997 and early 1998. His problems arose out of his inability to institute the reforms necessary to modernise the German economy and lower its stubbornly high level of unemployment. His decision in April to run for an unprecedented fifth term did not substantially improve his standing.

The Social Democratic Party (SPD) struggled with the question of its own candidate for chancellor for almost a full year, waiting until Gerhard Schröder won a handsome election victory in his home state of Lower Saxony in March 1998 before giving him the nomination. Following this emphatic endorsement, his rival, Oskar Lafontaine, the SPD Chairman and darling of the left, stepped back to allow the telegenic Schröder, who enjoys significantly greater popularity in the opinion polls than either Lafontaine or Kohl, to become the party's nominee. Taking a cue from New Labour in the UK, Schröder succeeded in impressing a more moderate, modernising cast on the party's programme. He has even been able to move the party toward accepting out-of-area combat operations by the Bundeswehr.

Alongside the struggle to achieve domestic reform, a new vision for Europe was taking shape in Germany, a vision that left behind the concept of a federal Europe in favour of a widening Europe with differing levels of development. Germany's own struggle to meet the Maastricht deficit criteria limited its ability to oppose Monetary Union membership for Italy, Spain and Portugal. With these countries part of the mix, EMU members would not be confined to the hard core most Germans had originally expected. Nor was Franco-German cooperation the vanguard it had once been.

Germany now found that on the question of enlargement, as on many other issues, it was closer to the UK than to France. To the surprise of many, Helmut Kohl stood in the way of allowing more decisions by majority vote, particularly on asylum and immigration, when the questions arose in Amsterdam. Germany's disproportionately high net contributions to the EU budget (60% of the total) were also increasingly questioned at home. In short, Germany's more differentiated approach to integration could no longer be so easily subsumed under a Eurofederal project nor driven by a Franco-German motor. There was still considerable symbolic value in the old concept, however, and in that sense Franco-German cooperation received a boost when Kohl and Russian President Boris Yeltsin announced in November 1997 that annual Franco-German–Russian summits would begin in the summer of 1998.

Germany took on a considerable role in the Bosnian peace process in 1997 and 1998, fielding the third largest contingent of troops and expressing willingness to help cover for US troop reductions after the end of the Stabilisation Force (SFOR) mandate. Domestic support for Bundeswehr participation in SFOR remained high, thanks to efforts made by Defence Minister Volker Rühe. Festering problems with right-wing extremists in the German military tarnished the Bundeswehr's image at home, however. On the question of military action against Iraq, Germany adopted a position closer to that of the UK than to the French. It backed the military option with political support, although it carefully avoided the question of contributing troops.

France Moves To Cohabitation

In April 1997, President Chirac called for early elections in France – a major miscalculation that left him at the head of a divided government. When he dissolved the National Assembly, Chirac aimed to solidify his parliamentary majority so he could secure further budget cuts and firm up the government's commitment to EMU. But enduring unemployment and the unpopularity of Prime Minister Alain Juppé led French voters to reject Chirac's appeal. Instead, in the election on 1 June, they gave Lionel Jospin's Socialist Party, operating in coalition with the Communist Party and the Greens, a majority of the parliamentary seats.

Despite early worries, France's latest experiment with cohabitation is working unexpectedly well, with Chirac and Jospin both enjoying high opinion poll ratings. Jospin moved adroitly from overly ambitious election promises to pragmatic political leadership. His initial pledge to create 700,000 jobs has been tempered by a clear commitment to fiscal austerity, but he continued to push for a 35 hour working week, a cause of concern to the country's business leaders. Balancing a social agenda with fiscal austerity was made more difficult by paralysing strikes in November and December, and adhering to the Maastricht criteria will not be easy. Still, the Jospin government demonstrated a commitment to Europe that few had expected.

In part, this is due to the moderate, pro-European foreign policy team that Jospin put in place, with Hubert Vedrine in the foreign ministry, Pierre Moscovici for Europe, and Dominque Strauss-Kahn in finance. Nevertheless, Jospin got into an early tussle with Helmut Kohl over an unsuccessful bid at the June 1997 Franco-German summit in Poitiers to add an employment clause to the EMU stability pact, as well as to create a political counterweight to the European Central Bank (ECB). The Jospin government also pushed hard to bring Italy and Spain into the monetary union, an objective not particularly popular in Germany. In November, France again sought to leave its mark on monetary union by nominating the head of the Banque de France Jean-Claude Trichet to head the European Central Bank. Most other EU members, particularly Germany, preferred former Dutch central banker and current head of the ECB precursor, Wim Duisenberg. Critics accused France of threatening EMU stability by making the issue of the head of the Central Bank a question of national, rather than European, interest.

France continued to differ with the US over the question of whether a European or an American should command NATO's southern flank, although this appeared to be a losing battle. Nor did Paris have any greater success in realising its desire to bring Romania and Slovenia into NATO in the first round of enlargement. France also diverged from the US on Iraq in late 1997 and early 1998, arguing against military strikes and for a more diplomatic approach. In part, Paris was motivated by what it saw as a special French role in the Arab world. But France was also poised to take advantage of commercial opportunities that would come with the lifting of economic sanctions against Iraq. Together with Russia, France was wary of excessive US influence, and French leaders in particular, repeatedly called for a 'multipolar' world.

The UK Turns To New Labour

On 1 May 1997, Tony Blair's 'New' Labour Party won a landslide victory, ending 18 years of Conservative rule by securing a 179-seat majority in the House of Commons. Blair's victory demonstrated that economics is not everything in electoral politics. In this case the incumbents lost despite a long period of economic growth and low unemployment (6.1% in 1997). It was Blair's success at transforming the old ideology-bound Labour Party, and his appeal as something new that helped give Labour its stunning victory. Even more important was the people's feeling that former Prime Minister John Major's government had become a weak and tired one, rent by scandals and deeply divided over Europe. In the end the Conservative Party headed a minority administration dependent on support from the Ulster Unionists, and it lost much of even its traditional support.

Blair's popularity held up remarkably well. Almost a year after his election he was still enjoying support in the polls of around 70%. Labour had found a successful blend of political innovation on constitutional reform and Europe, and 'tough love' on social issues. Blair pushed through welfare cuts in December, despite the opposition of 47 Labour members of parliament. On constitutional reform, Labour took a radical turn, initiating a devolution of London's power to Scotland and Wales. Referendums in September in the two regions will lead to the establishment of regional parliaments.

A new balance between reluctance to fully endorse the continental vision of Europe and a willingness to cooperate more closely with Brussels characterised Labour's policy towards Europe. The new Foreign Secretary Robin Cook, arguing that the UK intended to be a 'leading player' in Europe, indicated that the government would give up a number of obstructive positions taken by the Tories. It endorsed the EU's Social Protocol, accepting its inclusion in the Amsterdam Treaty, and it did not block inclusion of the Schengen Agreement on border controls in the treaty. On certain issues, however, the UK continued to demonstrate considerable opposition to further EU integration. It pressed hard for extending the time before competencies in the area of justice and home affairs would be 'communitised' under the Union's 'first pillar'. The same reluctance was shown when the UK joined with the EU's neutral states to prevent the Western European Union (WEU) from being merged into the EU.

On EMU, the new Labour government moved a step closer, but only a step. Gordon Brown, the Chancellor of the Exchequer, surprised many when, in his early days in office, he gave the UK central bank greater independence, ending the Treasury's traditional lead in the setting of interest rates. Nevertheless, like the Conservative government, Labour will not push for UK entry in 1999, or indeed during the current parliamentary session, as Brown made clear before Parliament in October 1997. Labour intends to take a 'wait and see' approach, making sure that EMU is a success before it considers joining and calling for a referendum to ensure that the people are behind the decision. In all of this was a certain irony: although the UK would not be joining EMU in the first instance, it would be presiding over the EU during the first half of 1998 when the EU would be deciding on monetary union.

The Blair government also moved closer to the US. Meeting in London in May 1997 and Washington in February 1998, Blair and President Bill Clinton charted a tougher stand on Bosnian war criminals and Republika Srpska, and confronted Iraq with a solid front on joint military strikes if it did not stop obstructing UN arms inspections. It was not aligned with the US on all issues, however. At the December 1997 Kyoto conference on world climate, the UK parted company with the US, working closely with

Germany to push (unsuccessfully) for a treaty reducing greenhouse gas emissions by 2010 to 20% below the 1990 level.

The UK was also occupied by remnants of the colonial era, including a final departure from Hong Kong as sovereignty shifted to China on 1 July 1997, political troubles in Montserrat which boiled up in the summer (triggered by a volcanic eruption), and a final resolution of access rights to Gibraltar in the run-up to NATO's Madrid summit in July 1997. Blair's greater engagement with both Europe and the US was paralleled by greater rhetorical reference to 'ethical foreign policy'. Yet it was engagement, not ethics, that stood out during Blair's first year at the helm of UK foreign policy.

A Green Light For Monetary Union

That Economic and Monetary Union would come into being in 1999 as planned had already seemed likely in the early part of 1997; that EMU would begin with 11 countries had not. But solid economic growth in Europe in 1997 eased the challenge of meeting Maastricht's criteria and led the Commission to recommend on 26 March 1998 that all EU members should join in the first round except the UK and Denmark (which wanted to stay out) and Greece and Sweden (which were not ready to come in). The disagreement that had erupted in June 1997 between France and Germany over EMU's political dispensation had been put aside. Controversy between those that planned to be among the ins and those determined to remain out, over the degree to which members could coordinate economic decision making in a 'Euro-X' body, was papered over at the December Luxembourg summit.

That the euro would come was clear, what it would bring was not. Putting Europe's diverse economies under one monetary policy would certainly force some countries to make significant adjustments, particularly in terms of product and labour market flexibility. Monetary union would no doubt accelerate the pace of eventual political union, but at the same time it could create tensions within the zone, and between those in and out of it, which may cause problems for further integration.

Revising The Treaty

The EU Treaty revisions, agreed in Amsterdam, underlined how difficult it could be for a Union of 15 states to find such unity of purpose. The most significant revisions came in the area of justice and home affairs, particularly with regard to bringing asylum and immigration policies under the EU's first pillar. A consensus was reached in Amsterdam on a five-year time period to communitise these policy areas, thereby also giving the European Commission, Parliament and Court a greater role. In this context, the Schengen Agreement to abolish border controls was drawn into the EU, although the UK and Ireland did not join in whole-heartedly. Instead they reserved the right to opt-in at a later date. Europol will also acquire an operational role during the five-year time period.

The EU also gained greater authority to decide on social policy with Blair's endorsement of the Social Charter and its inclusion in the Treaty. At Jospin's insistence, guidelines on employment were also attached to the Treaty, although Helmut Kohl blocked major new expenditure on job creation, insisting that employment policy in Europe should remain in national hands. On international trade, the summit leaders refused to give the Commission full power to negotiate agreements on services and intellectual property rights, leaving national governments significant control over these sensitive areas as well. The leaders only slightly expanded the areas subject to majority voting, among them research and customs cooperation.

Europe's weak response to the conflict in Bosnia had highlighted the shortcomings in the compromises that had been reached at Maastricht and there were hopes before the Amsterdam Summit that a tougher, tighter EU Common Foreign and Security Policy (CFSP) could be fashioned. Europe's governments, however, were not prepared to take the major steps on majority voting that would be required for such an outcome, nor were they prepared to create a distinct EU defence policy based on merging the WEU with the EU.

The EU did get a High Representative for its CFSP, but he was not given the stature that the French, in particular, had hoped. Instead, the post was given to the Secretary-General of the EU Council of Ministers. This relatively unknown, though important office was held in 1998 by a German, Jürgen Trumpf. The treaty revisions also established a planning cell for early warning and analysis, with staff drawn from the General Secretariat of the Council, the member states, the Commission and the WEU, to assist the High Representative in his work and to prepare decisions for the Council.

With regard to a common defence policy, Amsterdam brought the WEU's Petersburg tasks on crisis management under the roof of the CFSP, giving the European Council the power to set guidelines for their implementation by the WEU (revised article J.7). A protocol on enhancing cooperation between the WEU and the EU, including the introduction of parallel presidencies, was attached to the Treaty as well. A proposal by Germany and France, supported by Belgium, Greece, Italy, Luxembourg and Spain, to embark on a three-phase initiative to merge the WEU into the EU was left a possibility should the European Council so decide. This remained only a possibility, for the UK, Austria, Finland and Sweden were strongly opposed.

A new instrument, the 'common strategy' (a revision of articles J.2 and J.3) was created in Amsterdam. In limited cases its implementation could be decided by majority vote. The option for some countries to abstain constructively from participation in EU actions without actually blocking them was anchored into the treaty as well.

Amsterdam put off the difficult decisions required to streamline the EU's institutions and make them compatible with a membership that was expected soon to grow from 16 to 20 or even 25 countries. In a protocol, the members did make a commitment to pursue an arrangement whereby the commissioners would be reduced to only one per country, but on the condition that voting weights would shift to favour the larger countries. Before enlargement to more than 20 members, the EU would have to convene a conference to consider these reforms.

The idea of flexibility, which would allow some countries to adopt common policies, even if others choose not to, acquired a place in the revised treaty. But its application would be limited to those cases where there was unanimous agreement that flexibility could apply; clearly it would be limited to cases in which no country saw its vital interest threatened. The Amsterdam Treaty revisions were signed in October 1997, but the process of ratification, including a planned referendum in Denmark, would last well into 1998.

On The Way To Enlargement

Never before has the European Community or Union planned to bring in so many new members, with such different economies and political systems, in so little time. Enlargement means major reforms for the new members, but also for the old ones. The Commission's Agenda 2000, issued in July 1997, gave an idea of the magnitude of the change required. The Agenda made recommendations on the difficult task of writing a new EU budget to replace the one ending in 1999. It also proposed reforms of the Common Agricultural Policy and the cohesion funds, so as to keep overall EU expenditures below the ceiling of 1.27% of the EU's Gross Domestic Product (GDP), even in the face of the extra costs that enlargement would entail.

Attached to Agenda 2000, were the Commission's recommendations on the countries most ready for membership. The Agenda measures the applicants against criteria established in Copenhagen in 1993, including stable democratic institutions, a functioning market economy and an ability to accept and live up to the rules of the European Union. In a subtle ranking, the Commission's Agenda 2000 identified Poland, the Czech Republic, Hungary, Slovenia, and Estonia together with Cyprus, as ready to begin negotiations on accession. The remaining five new applicants, Latvia, Lithuania, Slovakia, Bulgaria and Romania will have to wait, as will Turkey, which has been an applicant since 1964.

In December 1997, the Luxembourg European Council endorsed the Commission's findings, and decided to begin negotiations with the six suggested candidates. All 11 Central and Eastern European applicants will negotiate 'pre-accession partnerships' that will pave the way for financial support and annual reviews of their progress toward eligibility. In March 1998, at a conference held in London which was attended by all the accession candidates, the EU formally opened enlargement negotiations. The mood was upbeat, although it was clear that accession talks will be long and difficult. With some 80,000 pages of European guidelines and regulations to be negotiated, it is unlikely that full membership will be achieved much before 2005.

The EU's approach to enlargement left long-time applicant Turkey particularly frustrated. Prime Minister Mesut Yilmaz reacted sharply to the Luxembourg decision to put Turkey in a special category, excluding it from consideration in either the first or second round, while including Cyprus (and threatening to include only the Greek part of Cyprus if the Turkish-occupied area refused to join). He temporarily ended contact with the EU and boycotted the London conference. Turkey's hostility was directed at Germany, in particular, which left German–Turkish relations in a bad state of disrepair. Germany is wary of Turkish membership because it fears that its large Turkish population would be likely to expand rapidly with the freedom of movement mandated by the EU's Single Market.

A Long Way To Go

Despite all the efforts devoted to revision of the EU Treaty and to enlargement questions, Europe still found it almost impossible to take a common stand on pressing international issues. Europeans criticised US unilateralist inclinations, but had little of substance to contribute themselves. When the revisions worked out in Amsterdam come into force, it is possible that this lack of cohesion will change. In the meantime, there were only a few common actions taken. In April 1997, when a German court linked the Iranian secret service to the killing of Kurdish opposition leaders in Berlin, the EU withdrew diplomats from Tehran and put its 'critical dialogue' on hold. In Bosnia, the EU played a significant role in reconstruction. In February 1998, the EU sent a fact-finding mission to Algeria. Yet, even in these areas, it was difficult for Europe to assert its identity and make a real impact.

Nor was the WEU significantly involved in crisis management. In a ministerial meeting in Paris in May 1997 it did agree to establish a military committee to meet twice a year. A number of commands were added to its Forces Answerable to the WEU list, and it did deploy a police-training force to Albania. Yet the reality was that European forces continued to shrink and that financial, if not political, constraints limited the prospect for any large combined arms capability. European assertion in the area of military crisis management still remained a distant prospect.

Indeed, it was defence economics as much as anything that was driving the creation of more interoperable European forces. Seriously challenged by the successful consolidation of the US defence industry and declining military budgets at home, European arms manufacturers and their governmental customers were slowly taking steps toward a greater integration of armaments production and harmonisation of procurement plans. The Eurofighter 2000 project was cleared to begin production when Germany approved the acquisition of 180 aircraft. The UK, Spain and Italy are the other partners in what will be Europe's largest collaborative arms effort. The EU Commission, aiming to protect Airbus turf, fought a tough fight against the merger between Boeing and McDonnell-Douglas in July, winning key concessions before approving the plan. And, in March 1998, high hopes were expressed that the UK, France and Germany would soon sign a memorandum organising closer collaboration among their aerospace industries.

While European governments made little headway during 1997 in their efforts to achieve a more unified and cohesive foreign and security policy with a new institutional basis, in other areas there were encouraging developments. The election in both the UK and France of moderate socialist leaders who espouse centrist positions, and the possibility of a similar result from the September elections in Germany, may help provide an atmosphere in which there will be more commonly shared views than there have been up to now. The successful agreements on enlarging NATO, membership of EMU, and future enlargement of the EU provide a new, and more solid, institutional basis for future European integration. In that regard, if Europe took one small step backwards in 1997, it took two significant steps toward that long-sought after, and elusive, goal. The real crunch will come in 1999 with the launch of the euro, whose success would ultimately force a more federalist approach, and whose difficulties or outright failure would plunge Europe into the doldrums.

◆

Leashing Ulster's Dogs Of War

For much of 1997 and early 1998, Northern Ireland hovered uneasily and precariously between war and peace. It was a year of recurring political and security crises, of excitement and despondency, of breakthroughs and setbacks. It was never quite war, at least not on the scale Northern Ireland had been used to; but nor was it ever quite peace, for the killings never stopped. The troubles claimed another score of lives.

There were four key events during the year. The first of these was the election in May 1997 of the UK's new Labour government. In July came the confrontation between Protestants and Catholics over the annual Orangemen's march through the town of Drumcree. The better than expected outcome raised hopes, and these appeared well founded when, soon afterwards, the Irish Republican Army (IRA) resumed its cease-fire. But in December a new outbreak of violence was sparked by the Republican assassination of Loyalist Billy Wright in the Maze prison. All these events had a huge impact which has not yet been fully played out.

Meeting, But Not Really Talking

Labour's election victory on 1 May 1997 transformed a fitful and unfocused political scene in Northern Ireland. Political talks had been convened at Stormont near Belfast on 10 June 1996, but they had become bogged down in procedural issues and no real negotiation had taken place. The talks at that stage included all the major Protestant Unionist parties, ranging from the biggest, David Trimble's Ulster Unionist Party, to the Revd Ian Paisley's Democratic Unionist Party and to two small but important groupings, the Progressive Unionist Party (PUP) and the Ulster Democratic Party (UDP). These two 'para-political' parties were important because they spoke for the two main Loyalist paramilitary organisations: the PUP for the Ulster Volunteer Force and the UDP for the Ulster Defence Association (UDA), which also goes under the name of the Ulster Freedom Fighters (UFF).

On the other side, the Constitutional Nationalists, which in Ireland means those elements who aspire to achieve a united Ireland eventually but disapprove of using violence towards that end, were represented by the Irish government and John Hume's Social Democratic and Labour Party (SDLP). But Irish Republicans were absent. Both London and Dublin insisted that, unless and until the IRA declared a cease-fire, its political wing, *Sinn Fein*, would not be admitted to the talks. An IRA ceasefire had been called on 31 August 1994 but it broke down in February 1996 amid Republican complaints that the then Prime Minister John Major had not dealt fairly with them.

In late 1996 and early 1997 the Republicans made fresh advances to the Major government, holding out the prospect of a renewed cease-fire if London would agree to certain terms. One of the most important of these was that John Major should drop the requirement for the IRA to surrender some of its weaponry in advance of talks. The two sides circled each other warily but in the event no business was done.

Part of the reason for Major's caution was the fact that a number of his back-benchers and, more importantly, some of his senior ministers, were against any attempt to strike a deal with the Republicans. His room for manoeuvre was further inhibited during the dying months of his administration by the fact that, by early 1997, his parliamentary majority had dwindled almost to zero. In this situation David Trimble's group of Ulster Unionist MPs assumed crucial importance for the Tories, since they held the power to save, or scupper, the Major government in tight votes. Major took note of the arithmetical realities and did little to offend Trimble who, Irish nationalists claimed, thus wielded inordinate influence in Whitehall and Westminster.

Meanwhile IRA violence went on. A soldier, Stephen Restorick, was killed by a sniper on 12 February 1997, while in England the IRA used bombs and hoax bombs to disrupt much of the motorway system, causing the loss of millions of pounds through their actions. One hoax led to the postponement of the Grand National, one of the world's great horse races, from its scheduled date of 5 April. However, the authorities privately acknowledged that the IRA was capable of bringing about a much higher level of death and destruction than this, and was in effect fighting only half a war – marking time, it seemed, until the long-expected general election.

A Turning Of The Tide

When the election arrived on 1 May and swept the Labour Party to power, one of the new Prime Minister Tony Blair's first acts was to launch an initiative to break the impasse in Northern Ireland. Overturning the previous government's policy, he authorised senior officials to talk directly to *Sinn Fein*. He also despatched to Belfast, as the new Northern Ireland Secretary, Dr Marjorie 'Mo' Mowlam, who quickly established a reputation for informality, enthusiasm and energy.

Contacts went on for some weeks, but there was a severe setback when the IRA shot dead two police officers in the County Armagh town of Lurgan on 16 June. For a moment it looked as though there would be no new IRA cease-fire, and that the hopes of a breakthrough which had been roused by the election result represented just another of the false dawns which have proved so dispiriting in Northern Ireland. Those hopes dwindled even further with the arrival of the 'marching season' – when Protestant organisations engage in their ostentatious annual parades. A particularly likely flash-point was Drumcree, County Armagh, where, in a by now traditional and fiercely-contested trial of strength, members of the Orange Order make an annual attempt to walk through a district which was once Protestant but is today Catholic. Exhaustive efforts to find a negotiated settlement failed and thousands of police were deployed to shepherd the march through the disputed area, and to hem in protesting Catholic residents.

This marked the end of Mowlam's honeymoon with nationalists, who bitterly accused her of giving in to pressure and the *force majeure* of extreme Protestants. The chances of a cease-fire seemed lost and the peak of the marching season, 12 July, seemed set to erupt into serious disorder. Then, against all expectations and almost all precedent, some sections of marchers withdrew their demands to march through Catholic areas and, in what seemed a near-miraculous deliverance, the day passed off peacefully.

This was followed by an even greater surprise when, in the third week of July, the IRA suddenly announced another cessation of violence. There was little of the ceremony and near-euphoria which had greeted the 1994 cease-fire; the eventual collapse of that enterprise had left almost everyone sadder, wiser, more suspicious and sceptical. It was nonetheless an historic moment in that it led the Republicans and London to fashion a deal to allow *Sinn Fein* into all-party talks for the first time ever.

The Blair administration had signalled that, in exchange for a cease-fire, *Sinn Fein* would, after a six-week waiting period, be granted entry into the Stormont talks. The IRA would not be required to hand over guns in advance of entry, and London further promised that the talks would be time-limited to prevent Unionist filibustering. While most elements welcomed the new cease-fire, it posed serious problems for Unionists, who had historically opposed negotiating with, or even speaking to, *Sinn Fein* or the IRA. Many Unionist politicians characterised the cease-fire as a disruptive stratagem rather than a genuine move in the direction of peace. If opinion polls were anything to go by, however, many grass-roots Protestants favoured giving peace a chance.

Paisley stormed out of the Stormont talks in October, declaring that he would never sit down with Republicans, but the two Loyalist para-political groupings stayed on, leaving Trimble's Ulster Unionists with a dilemma. They recoiled from the very idea of being in the same building as *Sinn Fein*, whom they accused of being unreconstructed terrorists, but, at the same time, Blair kept up a steady pressure on Unionists to talk. Unlike his predecessor he had an impregnable majority in the House of Commons and all the appearances of being in power for a decade. Political Unionism, which in the 1960s was commonly referred to as a monolith, had fragmented into five separate parties while opinion within its grass-roots ranged from the warlike to the pacific. A majority, however, appeared to be in favour of dialogue, if only because all else seemed to have failed. Eventually the Ulster Unionists faced up to the *Realpolitik* of the situation and, after many feints and much indecision, walked into the same building as *Sinn Fein*.

Paisley's indignant departure had demoted the talks from all-party to multi-party, but the UK and Irish governments persevered in his absence. Progress was slow, however. Although Unionists were often in the same room as *Sinn Fein*, the former resolutely refused to speak to the latter, either across the table, in the corridors or even in the men's room. And all through the autumn the arguments continued to be on formulating an agenda rather than substantive issues. The hope was that this tedious and apparently unproductive phase would turn out to have been a pre-negotiation period of some eventual value, and not simply a waste of time.

Outside the talks themselves, however, there was movement. Although Trimble would not talk to *Sinn Fein* President Gerry Adams, he had no objection to contacts with Bertie Ahern, who since spring had been the *taoiseach* (prime minister) of the Irish Republic. Despite some hiccups the two leaders appeared to have established a basis for a reasonable working relationship. Blair, meanwhile, opened personal contact with Adams, first meeting him in Belfast on 14 October and two months later, on 11 December, receiving a large *Sinn Fein* delegation in Downing Street. In Irish terms this was the stuff of history, since the last meeting between a UK prime minister and *Sinn Fein* had taken place in 1921 when Lloyd George met Eamonn de Valera and Michael Collins. It was certainly a much happier occasion than the last time Republicans made their presence felt in Downing Street, when an IRA mortar bomb exploded in the garden of Number 10 in 1991 and shook John Major's windows.

The visit was a welcome publicity boost to Adams, whose enthusiasm for the peace process had led to criticism within the IRA and *Sinn Fein*. In November a number of their members resigned, but, as time went by, it became evident that reports of a large-scale split had been much exaggerated. As 1997 drew to a close the process seemed in reasonable shape, and most of the end-of-year assessments were upbeat.

A Swing Back To Violence

It was at this moment that a major crisis was unexpectedly sparked off by the Republican assassination on 27 December of a leading Loyalist, Billy Wright. Wright, who had founded a breakaway group opposed to the peace process, the Loyalist Volunteer Force (LVF), was reputed to have been personally involved in a dozen or more Catholic killings. His record meant he had become an icon of Loyalist violence and the focus of hatred and fear among Catholics. He was serving a jail sentence when he was shot dead inside the Maze prison near Belfast. His killers were from the Irish National Liberation Army (INLA), a grandiosely-titled but fairly small Republican group which was also opposed to talking about peace.

Loyalist groups, including some which were supposed to be observing the cease-fire, had carried out sporadic killings throughout the year, often discreetly refraining from claiming responsibility for them. But the death of Wright, and of another Loyalist also shot by the INLA, were the signals for a killing-spree, with Loyalists retaliating by murdering eight Catholic civilians within a one month period. The killings were indiscriminate except in the sense that all the victims were chosen because of their religion. The series of deaths meant that tension soared, with many people in Catholic areas avoiding pubs and other potential targets and hurrying home in the evenings in a re-run of the bad old days of the worst of the troubles.

The rash of killings had deep implications for the overall situation. Neither the INLA nor the LVF, which carried out some of the killings of Catholics, had ever been involved in the peace process, but it emerged that three of the shootings had been carried out by the UDA. This meant that UDA gunmen had been active while their political representatives, the UDP, sat at the conference table. This posed a huge dilemma for other participants in the talks. Expelling the UDP would make it plain that those involved in violence could not also be involved in the political processes, which many viewed as an important point of democratic principle. But ejecting them would severely damage the prospects of arriving at an agreed settlement which the UDA would endorse, thus making the process much less inclusive than it had been. A compromise was effected through which the UDP was suspended from the talks with the promise that its participation would be reviewed if it could be demonstrated that their paramilitary wing was once again observing the cease-fire. In mid-February the Royal Ulster Constabulary determined that the IRA had carried out two killings the week before and the UK and Irish governments faced the more difficult question of how to handle a major and essential participant who appeared to have breached the basic principles on which the talks were founded. Although *Sinn Fein* leaders argued that it was unfair to hold them responsible for IRA actions, they were suspended for two weeks.

Before these actions the UK and Irish governments had issued their 'heads of government agreement' on an agenda for the peace talks. The agreement visualises a settlement based on three elements: an elected assembly that will provide government in Northern Ireland; cross-border bodies that will arrange links between the Irish government and the assembly in Ulster; and an indeterminate 'council' which would be made up of representatives from London, Dublin and the devolved assemblies in Northern Ireland, Scotland and Wales.

Sinn Fein leaders refused to return to the table when their two weeks suspension ended in mid-March 1998 until Blair agreed to talk once again with Gerry Adams at Downing Street. At the conclusion of that session, the Prime Minister proclaimed a settlement to be 'agonisingly close'. The US weighed in as well in an effort to boost the process; in addition to inviting all the leaders in Northern Ireland to St Patrick's Day celebrations in the US, President Clinton suggested that he might visit Belfast in May if an agreement were reached by then. Despite all the optimistic talk, however, the negotiations, which *Sinn Fein* rejoined on 23 March, were still balanced on a knife's edge.

Can They Get There From Here?

The difficulties in the way of an ultimately successful settlement are legion, and they come from both inside and outside the process. To begin with, the more the talks look likely to produce a positive outcome, the more likely they are to be violently attacked by paramilitary groups such as the LVF and the INLA who oppose all thought of compromise. There are at least three of these on the Republican side, all of which will be determined to help ensure that the process does not work. They have already drawn the lesson that violence can poison the general atmosphere and send tremors through the conference chamber; they will be trying to do so again.

Next, there is the external political threat. Paisley, who is completely opposed to the process, has wrecked many a previous initiative with his formidable destructive powers and he is poised to do so again. Anything like an agreement will have to survive a frontal Paisley assault, complete with the familiar but potent cries that Northern Ireland and Protestantism are in peril.

Then there are the tensions within the groupings involved in the talks. Almost all of them, from the Ulster Unionists to *Sinn Fein*, contain elements holding deep reservations about the whole process. Some Unionist MPs want the party to withdraw from the talks; some in the ranks of *Sinn Fein*, and the IRA, find it difficult to envisage how the process can lead to their traditional goal of a united Ireland. And, within the talks, the leaders of Unionism and Republicanism have emphatically not developed the kind of trust and goodwill which many believe necessary for a successful outcome. This reflects the state of affairs in the province at large. Community relations are at an all-time low and segregation is rife: many Catholics and Protestants prefer to live in places where they have no sight of each other, embittered as they are by so many years of communal conflict.

Yet this last superficially distressing point may embody one of the keys to a solution. One of the certainties to emerge from almost three decades of discord is that victory for any side has proven impossible. The IRA could not defeat the UK, nor crush Unionism; but then neither the UK nor Unionism could vanquish the IRA. That lesson has been absorbed, not by everyone but by almost everyone; and the logic that flows inexorably from it is that only a settlement offering an honourable compromise stands any chance of success.

Other factors are also of value in the search for peace. Blair is a UK prime minister with both parliamentary strength and personal determination to make progress in Northern Ireland. Furthermore, both the London and Dublin establishments have gradually come to believe that they share an aim, which is not territorial advancement, but containment and joint management of a common concern. This firm London–Dublin consensus represents a good reason for keeping alive the hope that, in spite of all the odds, substantial progress is a real possibility.

◆

Russia: A Semblance Of Order

'Our land is great and abundant, but there is no order in it'. So wrote the eleventh century Russian chronicler Nestor. In the nineteenth century, the poet Aleksey Tolstoy made the unending search for 'order' the refrain of his satirical romp through Russian history, *Istoria Gosudarstva Rossiiskogo Ot Gostomysla Do Timasheva*. In this perspective, 1997 can be seen as just another year in Russia's long pursuit of a tantalising but elusive goal. A measure of political order had been established between 1995 and the beginning of 1997 through parliamentary, presidential and finally regional elections. An economic order that would bring burgeoning business activity into a legal framework was the primary task for 1997.

Still Surprising Changes...

Boris Yeltsin's virtual absence from the scene for over six months following his re-election as president in July 1996 – he underwent a heart operation followed by double pneumonia – meant that serious policy making did not get under way until his return in March 1997. His agenda, shaped largely by his then Head of Administration Anatoly Chubais, was strongly reformist. It covered continued tight control over the money supply; the restructuring of industry (principally the gas and oil monopolies, whose tax evasion accounted for much of the shortfall in government revenue and the consequent failure to pay wages and pensions to state sector workers); a new and fairer tax code; a realistic budget; and a reduction in blanket subsidies, particularly on housing.

As so often in Russia, it was the organisational and personnel changes, as much as the policy statements, that served as the earnest of Yeltsin's intentions. He dispensed with the organisation of government inherited from Soviet times, in which ministers shadowed – and lobbied for – individual branches of industry, and strengthened the powers of the Finance Ministry, giving it overall control of expenditure, including that of the Ministry of Defence. The reorganisation temporarily sidelined Yeltsin's loyal but pedestrian Prime Minister Viktor Chernomyrdin. Centre stage was taken by Chubais, transferred to the government with the powerful portfolio of First Deputy Prime Minister and Minister of Finance, and the young provincial leader Boris Nemtsov, also promoted to First Deputy Prime Minister and given what he himself described as a 'kamikaze' portfolio of social policy and supervision of the natural resource monopolies.

One year later, on 23 March 1998, Yeltsin gave the kaleidoscope another shake, dismissing both Chernomyrdin and Chubais, and, formally, all the rest of the government as well. His stated reason, that people were not feeling changes for the better and that the government was lacking in dynamism, was manifestly inadequate for such a disruptive move and for placing the government in the hands of the inexperienced technocrat, Sergei Kiriyenko. Yeltsin's real motives were probably to confound those who were writing him off after another bout of sickness and to respond to a growing worry in his entourage that Chernomyrdin was becoming too 'presidential' in his prime-ministerial post while failing to convince that he could win the presidency in open competition.

The substance of Yeltsin's March 1997 initiative still looks as if it was a step in the right direction, but, to no one's surprise, its implementation was patchy. Inflation was brought down to 11% – half of the figure for 1996. The long decline in industrial production was halted and a 0.4% growth in Gross Domestic Product (GDP), the first in five years, recorded. A start was made on bringing to book some of the natural monopolies, *Gazprom* in particular. But the budget deficit, at 8% of GDP, is still too large and revenue during the early months of 1998 made no dent in it – predictably, since the start of the year is traditionally a bad time for revenue collection. The authorities had to balance the books by holding back large amounts of expenditure. Deliberate under-reporting by enterprises (to avoid taxes) means that economic performance is probably better than the figures imply, but this assumption itself brings no comfort, as it implies that the most dynamic part of the economy is outside the system. Worse, any sense of impending order was undermined by the continuing trail of assassinations. As well as leaders of known criminal gangs, the victims included heads of charities (made vulnerable by the attractions of the tax advantages that charities enjoy) and hoteliers, testifying to the continuing criminal links of many parts of the legal economy.

The picture was further darkened from the middle of the year by internecine warfare among the political and financial clans which had successfully engineered the re-election of Yeltsin a year earlier. The privatisation on 26 July of 25% of the giant communications holding company *Svyazinvest* became a *cause célèbre*. Two of Yeltsin's most influential financial and media backers, Boris Berezovsky and Vladimir Gusinsky, thought they had secured a gentleman's agreement from Chubais that they were the favoured bidders. When the auction was won by a higher bid from a consortium led by the Chairman of *Oneksimbank* Vladimir Potanin, a former First Deputy Prime Minister and close associate of Chubais, Berezovsky and Gusinsky launched a campaign to discredit Chubais, focusing on his unwise acceptance of a $90,000 advance on a book that was never published. Chubais' eventual dismissal was sweet revenge for Berezovsky, who may well have had a hand in Yeltsin's decision.

...Yet Underlying Stability

The political ructions at the turn of the year were aggravated first by a sharp drop on the Russian stock market resulting from the Asian economic crisis, then by a fall in world oil prices, hitting Russia's principal source of foreign currency. But crises such as these no longer presage serious political instability in Russia. This is due to three factors. First, although Yeltsin allowed his hand to be guided at the beginning of 1997 by radical economic reformers, politically he was looking for compromise in 1997, the 80th Anniversary of the Revolution, which he designated a year of reconciliation. His opponents, for their own reasons, reciprocated. Yeltsin created an exclusive forum involving himself, then Prime Minister Chernomyrdin, the Chairman of the Federation Council (upper house) Yegor Stroev, who is also a prominent regional governor, and the Chairman of the Duma (lower house) Genady Seleznev, a leading Communist. This 'Council of Four' discussed and helped to defuse the autumn crises, at least in the context of relations between president and parliament. The opposition communist and Vladimir Zhirinovsky (nationalist) factions in the Duma, which together form a majority, proved again that their primary consideration was to preserve their seats against the ever-present threat of the Duma's dissolution. As a result, opposition motions of no confidence routinely failed, the 1998 budget has been approved after a delay, and the important new tax code stands a chance of being approved in the first half of 1998 and becoming the basis of the 1999 budget.

The second factor preserving political stability was the striking disinclination of the extra-parliamentary opposition to champion the cause of large, discontented sections of the workforce. The main trade union movement organised a day of protest on 27 March, but failed to convince either its supporters or the government that it has real mobilising power. True to its Soviet heritage, it is more interested in sharing power with the political establishment than opposing it. The ultra-left communists lacked the organisation to match their rhetoric, and General Alexander Lebed, once again in opposition following his brief period in government in late 1996, was reluctant to try to mobilise a force so fragmented and unpredictable.

The third reason the country has now proved better able to ride out political crises lies in the growing power of regional leaders. All are now elected, and this limits Moscow's ability to control them. But many regions still depend on subsidies from Moscow or on the country-wide infrastructure to transport and market the natural resources on their territories. All pragmatists, the regional leaders modulate their tactics to extract the maximum funds from Moscow, while demonstrating their power and influence locally. Such tactics often involve confrontation with the centre, but preclude attempts to undermine the system, even among those governors who are Yeltsin's enemies. (Governor Vasilii Starodubtsev of Tula was one of the anti-Gorbachev putchists of 1991, for which he served a prison term, and Governor Alexander Rutskoi of Kursk was formerly Yeltsin's vice-president and a leader of the parliamentary revolt against him in 1993, for which he too was jailed.)

Chechnya, the exception, is in limbo. There is not even the semblance of order there. Moscow does not rule, but, without recognition by the international community, Chechnya's claims to independence are hollow. During most of 1997 there was a circumspect search for compromise between the newly elected President Aslan Maskhadov, who had negotiated the peace agreement of August 1996 with General Lebed, and Lebed's cautious successor as Secretary of the Security Council, Ivan Rybkin. But by the end of the year, Maskhadov's moderate line was under strain domestically, and he ceded much of his power to the field commander Shamil Basaev, whom he appointed acting prime minister. The appointment is a severe public relations problem for the Russian authorities as it was Basaev who led the disastrous hostage-taking raid on a Russian hospital in 1995, for which he is on the wanted list in Russia. But Basaev, no less than Maskhadov, is driven by the desperate need to bring economic activity to his poverty-stricken republic, for which a measure of cooperation with Russia is a precondition. A good sign was the start of transport of Caspian oil to the Black Sea through Chechnya, although the Russians are prudently planning an alternative route that by-passes the republic. Despite the dissenting voice of the then Interior Minister Anatoli Kulikov, the man in Moscow nominally responsible for internal order in Chechnya, there is no disposition to resort to force against Chechnya. Rather, military units are being deployed round its perimeter in an attempt to prevent Chechen violence infecting the rest of the North Caucasus, particularly neighbouring Dagestan, with its significant Chechen minority.

Reforming The Military

The armed forces are in any case in no state to undertake any serious military activity, with draft-dodging rampant and morale at a low ebb. Even here, though, there are signs that a corner has been turned. During

1997 progress was made on the long awaited reform of the military. The arrival of new Russian Defence Minister, Igor Sergeyev, on 23 May helped to resolve a protracted stand-off between the Defence Council, under civilian leadership, and the Defence Ministry on the basic principles behind the changes. The fundamental concept of reform through to 2005 was finally approved in June by the President, giving the go-ahead to a more consistent effort on this front.

Neither the concept of reform, however, nor the concept of national security, another piece of paper approved by the President in December, provided an adequate strategic or financial foundation for the development of modern armed forces. Military reform in 1997 was focused on structural (primarily administrative) adjustments rather than numerical or more fundamental strategic ones. Within the space of six months the Russian Defence Ministry successfully completed a merger of the Strategic Missile Troops, the Space Missile Defence Troops and the military-technical troops, resulting in a cut of over 30% in the administrative personnel of these forces. The Ground Troops command structure was reorganised and a merger of the Air Force and Air Defence Force was practically completed by early 1998.

Despite their severe economic difficulties, the military have not engaged in any social protest, and, for the most part, they have stayed out of politics. It has become clear, however, that military reform will become one of the major issues in the next presidential election. Even in 1997, the issue was being actively targeted by various political parties and movements. In approving the 1998 budget on 10 February 1998, the Duma voted overwhelmingly for a 1% cut in each budget category, with the savings to go to military reform. The next, and most challenging, phase involves dropping another 200,000 men from the current total of 1.5 million by the end of 1998.

The Cost Of Stability

All this points to 1998 being another critical year, but one in which the emerging stability will probably not be challenged. There is a price to be paid for this, summed up by the terms currently used to describe the state of the country: 'democratic feudalism', meaning the right to choose your own patron but no more, and 'oligarchic capitalism'. The prerogative of an elected governor to control his fiefdom as he wishes, including the elected local council, is rarely questioned. The media is free in the sense of not being controlled by a single central authority as in Soviet days, but it is coming increasingly under the control of rival business and political interests. The failure of the fiercely independent editorial board of *Izvestiya* to stave off a commercial and political takeover was one of the setbacks of 1997. So was the eventual passage into law of a new bill on freedom of conscience and religious organisations that discriminated unashamedly in favour of the Russian Orthodox Church, now the established church in all but name.

Russia In The World

Russia's position in the world continues to decline in traditional geo-strategic terms, although its sheer size and potential allow it to wield an influence out of all proportion to its current economic and military strength. A growing self-confidence, coupled with the return of Yeltsin to the world stage, allowed Russia to make the best of the available openings in 1997. Yeltsin's summit meeting with US President Bill Clinton on 21 March set the scene for the signing on 27 May of the Founding Act on Mutual Relations, Cooperation and Security between Russia and NATO and the involvement of Russia in NATO organs, from the Permanent Joint Council down, that were specially created for it. Thus NATO enlargement did not stimulate the visceral anti-Western mood in Russia that it once threatened to do. Although the Strategic Arms Reduction Treaty (START II) remains unratified, the ratification of the Chemical Weapons Convention in November was noteworthy as the first arms control agreement to be ratified by Russia's new parliament. The NATO enlargement story is not, however, over. It is one thing for former members of the Warsaw Pact to join NATO and another for former republics of the Soviet Union to do so. Discussion of a further wave of enlargement, to include some Baltic states, will revive Russian sensitivities and ensure that 1998 is no more comfortable a year than 1997.

Despite its efforts, Russia was unable significantly to influence the course of events in Iraq during 1997, but its other diplomatic achievements were impressive. The list included the signing at the end of May of the long-awaited treaty with Ukraine and an agreement on the division of the Black Sea Fleet; a border agreement with Lithuania and visits to Moscow by the Lithuanian and Latvian presidents (the first since their countries gained independence); symbolic if not very substantial agreements with China on 'strategic partnership' and border demarcation; and a commitment by Yeltsin and Japanese Prime Minister Hashimoto at their summit on 2 November to sign a peace treaty by 2000. The signature of a Treaty of Union with Belarus on 2 April was a mixed blessing. It reflected a widespread desire in Russia to restore ties with the Slavic republics of the former Soviet Union, but was tarnished by the increasingly unsavoury nature of President Alexander Lukashenko's regime in Belarus.

Russia's increasing integration into the world economy in 1997 was illustrated, for better or worse, by the dip on the Russian stock and financial markets caused by the Asian economic crisis. Despite continuing doubts in international financial institutions over the course of reform (in

October the International Monetary Fund (IMF) temporarily withheld a tranche of its $10.1 billion loan), political imperatives ensured that in 1997 Russia was successful in its efforts to join international bodies. It was formally admitted to the Asia Pacific Economic Council, as well as the Paris and London Clubs, which puts Russia on something like an equal footing with other industrialised countries in international debt matters. Small steps were taken in Russia's increasingly close association with the Group of Seven Industrialised Nations. But perhaps the most dramatic development has been the involvement of major Western oil and gas companies in Russia following the lifting of the 15% limit on foreign stakeholding. Both Shell and British Petroleum signed strategic agreements with major Russian conglomerates in 1997. The expected privatisation of *Rosneft*, the only remaining energy-producing enterprise still wholly in state hands, will see the two oil giants pitted against each other. International participation should help to ensure that the auction is transparent and fair, along the lines of the *Svyazinvest* transaction and in sharp contrast to other recent government sell offs, which have become a byword for insider dealing among the top political and business clans.

Yeltsin And After

Much of the credit for Russia's enhanced position in the world goes to Yeltsin himself, despite his unpredictable behaviour when abroad. His interventions in the internal debates on NATO enlargement and the treaty with Ukraine were crucial.

Nonetheless, as his dismissal of the government demonstrates, his actions are now increasingly dominated by his impending departure from the scene. He repeatedly denies that he intends to run again for president. These denials cannot be taken as final, especially if the Constitutional Court rules that his current term is actually his first, not his second (his first term started in 1991 under the Soviet regime). Yet on current form it is hard to see Yeltsin as Russia's president beyond July 2000, and the small field of potential successors is already manoeuvring for position.

Polls indicate that one pattern of the 1996 race will be repeated. Gennadi Zyuganov, or whoever is leading the Communist Party at the time, should have sufficient bedrock support to get into the second round, but is unlikely to attract the extra votes needed to win. He will face the establishment's preferred candidate. There is currently a gap to be filled here unless Chernomyrdin can secure the financial and media backing to make a significant bid from outside the government. He could be challenged by Moscow mayor Yuri Luzhkov if the latter, by spreading his money widely enough, can convert his local importance into popularity at the national level. An outsider as the establishment candidate is Yegor Stroev, governor of the Orel region as well as number three in the current state hierarchy as leader of the upper house. The ground has been slipping away from General Alexander Lebed, who came third in the 1996 elections but has been unable to bolster his appeal with a solid organisational or political base. In the current rather conservative climate, representatives of the younger generation, such as First Deputy Prime Minister Nemtsov, would be wise to hold off till 2004.

Whatever the outcome of the next electoral round (the Duma elections in December 1999 precede the presidential race), elections are only one of the elements shaping the new political profile of Russia. The increasingly complex web of relations in politics, business and society, not least in their regional dimension, will determine whether Russia moves further towards a civic society rooted in the rule of law or tries to maintain a semblance of order based on feudal democracy and oligarchic capitalism.

The Skies Brighten Over Bosnia

Although the prospects for lasting peace in Bosnia continued to be fragile, there were grounds for greater optimism towards the end of 1997. In particular, the NATO Stabilisation Force (SFOR) and the Western-led Peace Implementation Council began to work more closely together. The principals involved carefully avoided using the term 'protectorate', but Western policy certainly grew more assertive once Madeleine Albright replaced Warren Christopher as US Secretary of State and Robin Cook replaced Malcolm Rifkind as British Foreign Secretary. Albright and Cook used much tougher rhetoric in addressing the Serbian, Croatian and Muslim leaders than their predecessors had ever done and were more committed to re-integrating and democratising Bosnia than accepting *de facto* partition.

Other new appointments, combined with new policies, also helped to make Western policy more active. In NATO, General Wesley Clarke replaced General John Joulwan as Supreme Allied Commander Europe. In Sarajevo, former Spanish Foreign Minister Carlos Westendorp replaced Carl Bildt as High Representative in charge of implementing the civilian aspects of the November 1995 General Framework Agreement for Peace in Bosnia and Herzegovina (the Dayton Accords) and Jacques Klein, who had successfully led the UN Transitional Authority in Eastern Slavonia (UNTAES), moved to Sarajevo as principal deputy to Westendorp in mid-1997. The Peace Implementation Council, at its meetings in Sintra in May and in Bonn in December, authorised Westendorp to impose the Framework Agreement in cases where the local parties were reluctant to accept and act on its principles. Westendorp used this authority to impose a common Bosnian currency and licence plates in January 1998, and a common flag in February.

Supporting Civilian Tasks

Despite fielding only half the manpower that the Implementation Force (IFOR) had at its disposal, SFOR was more supportive of the civilian aspects of the Dayton Accords in 1997 than IFOR had been in 1996. In actions coordinated by SFOR's Civil Military Task Force, NATO troops assisted with the repair and maintenance of bridges, roads and railways; opened up the Banja Luka, Tuzla and Mostar airports; and supported the Organisation for Security and Cooperation in Europe (OSCE) in its electoral and arms control tasks. It also organised joint patrols with the unarmed UN International Police Task Force, to remove excess weaponry from local police stations and to dismantle illegal roadblocks.

From SFOR's perspective this was probably regarded as dangerous mission creep, but for most Bosnian civilians it fell short of what they wanted. Although these and other actions were stipulated in an Annex to the Dayton Accords, NATO commanders were still reluctant to take on non-military tasks that might endanger their own personnel. There was, for example, little NATO help with clearing mines (except where troops were vulnerable); with law enforcement; with protection for returning refugees and displaced persons; and, until mid-1997, with the arrest of war criminals.

Jacques Klein's UNTAES troops, rather than NATO, broke the 18-month paralysis in bringing war criminals to justice. In late June 1997, they arrested Slavko Dokmanovic, a Croatian Serb indicted for the massacre in Vukovar in November 1991 of 200 non-Serb hospital patients. In early July, the Chief Prosecutor at the International Criminal Tribunal for Former Yugoslavia, Louise Arbour, praised the UN, and, suggesting that NATO was evading its responsibility, called SFOR 'an exercise in self defeat'. Whether they were piqued at these remarks or not, a British NATO team shortly thereafter arrested Milan Kovacevic and shot and killed Simo Drljaca as he resisted arrest. Both Kovacevic and Drljaca had been indicted for their role in the notorious Omarska camp in 1992. In October, economic pressure forced Croatian President Franjo Tudjman to send ten indicted Croats to the Hague. In December, in the British sector, Dutch commandos arrested Vlatko Kupreskic and Anto Furundzija, two Bosnian Croats indicted for the massacre of Muslims in the village of Ahmici in the Lasva Valley in April 1993. And in January 1998, a US-led snatch team arrested Goran Jelisican, a Bosnian Serb indicted for his role in the Serb-run Luka camp near Brcko. International Criminal Tribunal prosecutors welcomed these arrests, but Chief Prosecutor Louise Arbour deplored the immunity offered war criminals in the French sector and noted that, in addition, more than 50 of those who had already been indicted, including Radovan Karadzic and Radko Mladic, remained at liberty.

Re-merging Populations

Repatriation was painfully slow in 1997, in part because refugees did not want to return to communities still run by indicted war criminals, but also because there was no public order or freedom of movement. The problems were compounded by inadequate housing (housing from which one group of refugees was expelled was often occupied by other groups of refugees), poor job prospects, and resistance in many Bosnian communities to the return of minority groups. Of the 2.3 million people displaced by the war, by the end of 1997 only 381,000 had returned home (171,000 refugees and 210,000 displaced persons). In most of these cases, these were people returning to areas where they belonged to the ethnic majority. Only 22,500 (13,800 Croats, 5,600 Serbs and 2,900 Muslims) returned to areas in which the army of their ethnic group did not control the territory. The vast majority of these returned to the Muslim–Croat Federation, but 1,200 Muslims returned to the Zone of Separation on the Republika Srpska side of the Inter-Entity Boundary Line and some 1,100 Muslims have returned to villages close to Brcko.

In March 1997, the UN High Commission for Refugees (UNHCR) launched an ambitious open cities project. This initiative promised economic assistance not just to individual returnees, but throughout municipalities which were prepared to welcome refugees and displaced persons regardless of ethnicity. Four Federation cities were designated open cities in 1997: Konjic, Busovaca, Vogocsa and Bihac. A further 30 municipalities applied for this status, including some in Republika Srpska.

For open cities to work well the UNHCR needs NATO to guarantee freedom of movement, provide assurance that returnees can remain, and coordinate returns with housing-construction schedules and job opportunities. Other urgent requirements are to reform the property laws and create a competent and independent judiciary. During 1996 and 1997 returnees suffered serious difficulties from disputes about ownership which could not be fairly or quickly settled because most judges were dependent on and responsive to local party bosses.

The UNHCR also needs asylum countries to continue temporary protection until conditions improve sufficiently for it to be safe for minorities to return. Most asylum countries continue to cooperate, but in Germany (which took in some 350,000 Bosnian refugees during the war – more than half as many refugees as the rest of world combined) many local authorities ran out of patience (and/or funds) and started premature deportations in early 1997. This forced the UNHCR not only to relax its own standards (for example, that all returns be voluntary) but also to contravene Annex 7 of the Dayton Accords by encouraging the return of refugees only to those areas where they belonged to the ethnic majority. Many Bosnians argued that this was another form of ethnic cleansing, and it did little to further the goals of reconciliation and integration.

Key Ingredients

Returnees will have more incentive to return home when the economy improves and there is some prospect of gainful employment. The second year of the Priority Reconstruction and Recovery Programme for Bosnia and Herzegovina, which the international donor community endorsed at the end of 1995, was broadly on track at the end of 1997. At three pledging conferences in 1996–97, co-hosted by the EU Commission and the World Bank, donors committed over $3 billion for reconstruction and peace implementation, of which more than half was disbursed. The EU was the main donor, followed by the US and Japan. Within the EU, the Netherlands was the most generous, followed by Italy. Norway was the most generous of the non-EU donors in Europe. Donors were mindful of the need for better accountability as, inevitably, some funds were misused. As reported by the European Commission's Customs and Fiscal Assistance Office, and the World Bank's Public Expenditure Review, too much was disbursed to the *Mafiosi* as protection money and into other corrupt and nefarious activities.

The large bulk of the funds (97%) were disbursed to the Federation, due to its greater effort to comply with the Dayton Accords than Republika Srpska which continued to be obstructive for most of 1996 and 1997. According to the Federation Statistical Institute, the Federation economy grew by almost 68% in 1996 and by 40% in 1997 to reach approximately 47.5% of its pre-war level. Future aid disbursements were planned at 70% for the Federation and 30% to Republika Srpska based on the relative size of population and degree of physical destruction to be repaired.

Another missing prerequisite for a return to normality was pointed up at the December 1997 meeting of the Peace Implementation Council in Bonn, which deplored the fact that protection of human rights in Bosnia and Herzegovina was still far from adequate. Annex 6 of the Dayton Accords sets out an Agreement On Human Rights, in which the parties agree to secure for all persons within their jurisdiction 'the highest level of internationally recognised human rights and fundamental freedoms'. Annex 6 also established a Commission on Human Rights, a Human Rights Ombudsman and a Human Rights Chamber in which alleged violations could be addressed. Human rights lawyers working in Bosnia, however, report that little of this works, mainly because none of the many organisations which have a theoretical responsibility for human rights (for example, the OSCE, the International Police Task Force, the European Community Monitor Mission, the UN mission to Bosnia and Herzegovina and the Human Rights Co-ordinating Centre in the Office of the High Representative) takes the job seriously. There is plenty of reporting, but little analysis and hardly any follow-through in terms of investigation and punishment of abuse. Many critics have called for the appointment of a high level Commissioner with enough power to pull together and enforce the human rights brief.

New Leadership In Republika Srpska

From the beginning of the Yugoslav wars of dissolution, Western policy was particularly weak in searching out possible alternatives to the aggressive and virulently nationalistic leaders throughout the area. Instead, Western diplomacy used President Slobodan Milosevic of Serbia and President Franjo Tudjman of Croatia, even to the point of letting them help shape the final peace settlement. In 1997, however, some of this tolerance began to fade. In Bosnia, both NATO and the Office of the High Representative sought to isolate the hardline Serbs loyal to Radovan Karadzic and the Serb Democratic Party (SDS) in Pale, by supporting Biljana Plavsic's Serb Peoples' Alliance in Banja Luka. To be sure, Plavsic appeared on the surface to be an unappealing choice given that during the war she had been one of the most virulently anti-Muslim and aggressively nationalistic members of the Bosnian Serb leadership. In her favour, however, was that she was 'the enemy of our enemy'. With this in mind, in the second half of 1997, SFOR troops defended police loyal to her against those loyal to Karadzic and closed down a number of TV transmitters controlled by the SDS. Nevertheless, Plavsic, while paying lip-service to the Dayton Accords, continued to obstruct the return of non-Serbs to western Republika Srpska and refused to recognise the International Criminal Tribunal.

Despite NATO's help, the November 1997 elections turned out to be inconclusive. Karadzic's SDS party lost its majority in the parliament, but it retained enough votes to bloc Plavsic's first nominee for prime minister. Plavsic's second choice, a West-leaning democrat named Milorad Dodik, squeezed in by a single vote in January 1998, after most of the hardline deputies had gone home. Dodik's position is exceedingly fragile. He is dependent on the uncertain support of foreign aid and the votes of 16 sympathetic Muslim deputies. If he manages to survive, the prospects for positive developments in Republika Srpska will be vastly improved. Dodik was a successful businessman before the war, and was elected mayor of Laktasi, near Banja Luka, in 1987. In 1990 he was elected as a candidate for the party of Yugoslavia's last reformist prime minister, Ante Markovic (Alliance of Reformist Forces). During the war, Dodik formed an opposition block of anti-war deputies in the Bosnian Serb parliament but chose not to run for office in the 1996 elections. In the municipal elections of September 1997, however, Dodik's party won majorities in Laktasi and Srbac, after which Dodik immediately invited all those non-Serbs displaced from these two communities during the war to return home. Despite this exemplary gesture, Laktasi received no reconstruction aid from the international donor agencies.

In contrast, when he became prime minister Dodik received early endorsements and promises of aid from both Brian Atwood, Administrator of US AID, and from EU Commissioner Hans van den Broek. Dodik has been receiving support because – in stark contrast to the SDS – he espouses separation of church and state, a free press, and full support for the Dayton Accords, especially the safe return of all refugees and displaced persons. In his first two weeks in office, Dodik signed agreements to form multi-ethnic police forces, agreed to issue common passports and licence plates, agreed on a common currency, drew up a proper budget, sacked a host of corrupt officials, endorsed a plan to bring refugees home and began the process of transferring the seat of Republika Srpska's government from Pale to Banja Luka.

Unfinished Business: The Brcko Arbitration

The Dayton proximity talks in November 1995 failed to determine which entity should control Brcko, the pivotal city on the Sava River in north-east Bosnia which divides the two halves of Republika Srpska and also links the Federation with Croatia. The Dayton Accords provided for the issue to be settled by arbitration by 14 December 1996, but the presiding arbitrator (Roberts Owen) put off the decision twice (first until February 1997 then until March 1998). Between March 1997 and March 1998, Brcko was administered by Robert Farrand, a US diplomat.

The International Crisis Group, a non-governmental organisation which monitors the peace process in Bosnia, recommended that Brcko not be awarded to either entity, but to the common institutions of the federal state of Bosnia and Herzegovina. Until Dodik's election it would have been unthinkable to award Brcko to Republika Srpska, but Dodik says that he would not survive as prime minister if Brcko is awarded to the Federation. Westendorp has hinted that, with Dodik as prime minister, he could imagine awarding Brcko to Republika Srpska but this would be a hugely difficult decision for the Federation to swallow as it would imply condoning the expulsion of the majority non-Serb population during the war. Such a solution could only work if Dodik cooperates fully with NATO to bring indicted Bosnian Serb war criminals to justice.

The New Military Balance

European anxieties that the US programme to train and equip the Federation armies would undermine the post-Dayton arms control regime in former Yugoslavia eased somewhat in late 1997. Annex 1-B of the Dayton Accords provided for a regime of confidence and security building measures modelled on those of the OSCE; a sub-regional arms control agreement modelled on the Conventional Armed Forces in Europe (CFE) Treaty and, eventually, a Balkans-wide arms control regime.

The US train-and-equip programme stemmed from a promise to President Alija Izetbegovic in Dayton in November 1995. It is managed from Washington by James Pardew, who heads the US State Department's Task Force for Military Stabilisation in the Balkans, and is conducted in Bosnia by the US-based private sector military company, Military Professional Resources Inc. The European allies, and other military personnel serving with IFOR and SFOR, feared that a programme which enhanced the military forces of one entity (the Federation) would undermine the interest of the other (Republika Srpska) in complying with the sub-regional arms control agreement. There was also some anxiety that train-and-equip might embolden dissatisfied Muslims in the Federation to retrieve territory they had lost during the war by force of arms and at Dayton.

Defenders of train-and-equip emphasised that the programme only built up the Federation armies within the limits of the sub-regional agreement. This agreement set ratios for heavy equipment that give the Federation a two-to-one advantage over Republika Srpska (on the assumption that Serbia proper would come in on the side of the Serb entity in any future conflict). Equipping the Muslims and Croats in the Federation with more sophisticated weaponry than that of the Bosnian Serbs further exacerbates what seems to some an already unfair balance. At the same time the training component of train-and-equip has the more positive and restraining effect of instilling Western military discipline, and respect for humanitarian law, into what were undisciplined Muslim and Croat armies, both of which committed appalling atrocities (albeit mostly in response to attack) during the war.

As Europeans feared, the Bosnian Serbs did exploit all the loopholes in Article III of the sub-regional agreement, but when the period for the reduction of arms ended on 21 November 1997, all the parties – both Bosnian entities as well as Croatia and the Federal Republic of Yugoslavia (Serbia and Montenegro) – appeared to be in compliance, having destroyed 6,580 pieces of equipment. The OSCE and SFOR continue to monitor the parties during the residual validation period and a compliance report is due on 1 June 1998. In hope of bringing about in 1998 some improvement in the slow progress that has been made thus far in negotiations aimed at a wider Balkan arms control regime, the OSCE appointed a Special Representative to take charge of these negotiations.

Securing The Future

In December 1997, President Bill Clinton admitted that previous exit strategies had been a mistake and promised that US troops would stay in Bosnia as long as it took to stabilise the peace. This still begged the question of the level of the US contribution to the post-SFOR force and whether there would be a different mandate for NATO troops after June 1998.

Three options were under discussion at NATO's Military Committee in Brussels in early 1998: maintaining the same 30,000 troop stabilisation force; reducing to a 20,000-strong deterrence force; or cutting back to an even smaller tripwire force. Since most European countries were prepared to maintain the same number of troops that they were deploying for SFOR, the size of the new force will depend on decisions in Washington. Westendorp was adamant that any post-SFOR deployment be at least at the same strength as SFOR given the Dayton implementation tasks still to be accomplished. The US Defense Department wanted a smaller deterrent force, which would shift most of the stabilisation tasks to the Europeans. The White House and State Department, however, wanted US troops to lead on the ground until peace had stabilised, consistent with NATO's declared new mission to provide stability throughout post-communist Europe.

In addition to the number of troops, the question of their mandate was still to be decided. The Dayton Accords had made no provision for domestic law enforcement in Bosnia. As NATO Secretary-General Javier Solana said in November 1997: 'Experience in Bosnia has revealed a gap between the ability of SFOR to provide a security environment and the problems of domestic police forces in guaranteeing law and order under democratic control'. Solana proposed a UN standing police force to deal with the problem. In January 1998, US Secretary of Defense William Cohen proposed instead a battalion-sized paramilitary force under NATO command in each of the three SFOR sectors. The main function of the paramilitaries would be crowd control when the local police were unable or unwilling to cope. Unfortunately, there was no consensus within the Peace Implementation Council about how to deal with law enforcement in Bosnia (as there is not on how to manage law enforcement in peace support operations generally). Countries with armed gendarmeries like France, Spain and Italy were sympathetic to arming an international police presence in Bosnia, but the Nordic countries and the UK tended to oppose armed police.

And In The Neighbourhood...

There will not be a stable peace in Bosnia so long as all of Bosnia's neighbours are not also stable and at peace. One way to help inspire democratic behaviour in former communist countries is to dangle the carrot of membership in Western institutions, which implies that the EU, and preferably also the US, will need to commit themselves to remain in the Balkans for the long haul.

Of the former Yugoslav states, Slovenia is already on track to join the EU, and Croatia, which has its sights on the same goal, is making efforts to appear an effective prospective member. In October 1997, President Tudjman delivered most of Croatia's indicted war criminals to the Hague and, in January 1998, Eastern Slavonia passed from UN to Croatian control without incident. Notwithstanding these positive signs, Croatia has much further to go to meet EU criteria on pluralism, protection of human rights and freedom of the press. In the Federal Republic of Yugoslavia, developments in Montenegro were more encouraging after the election of Milo Djukanovic as president ousted Milosevic's protégé Momir Bulatovic and thus opened a gap between Podgorica and Belgrade. In Serbia, however, Milosevic's influence remains pervasive given that another of his protégés, Milan Milutinovic, won the presidency over the extremist Vojislav Seselj.

The most serious threat to peace in the area arises from the role of Albanians in the Balkans. Discrimination against ethnic Albanians in Kosovo is a constant source of concern, not least because they are losing patience with the passive non-confrontational tactics of Ibrahim Rugova and his Democratic League of Kosovo, and are paying more heed to the more radical Kosovo Liberation Army (UCK). Indeed, support for the UCK among ethnic Albanians in the province has increased following a full-scale military assault by Serbian forces on Drenica on 5 March 1997. Any further unrest could undermine Rugova's position and potentially lead to the spread of violence beyond Serbia's borders. In Macedonia, given the far from stable situation of the minority Albanian population, there is concern that it may be very difficult to preserve peace, particularly if the UN Preventive Deployment force departs in August 1998. At least in Albania itself, although it has not yet fully recovered after the violence which followed the collapse of the pyramid schemes, prospects are better under Prime Minister Fatos Nano. The economy is being restructured under the lash of the International Monetary Fund and the Western European Union is training and monitoring the local police.

Room For Cautious Optimism

The two most encouraging events for Bosnia as 1998 opened were President Clinton's announcement in mid-December 1997 that US troops would remain in Bosnia beyond the previously scheduled exit date of June 1998 and the election in mid-January 1998 of Milorad Dodik as prime minister of Republika Srpska. The prospect that US troops would stay on the ground boosts the confidence of all those engaged in the peace process and removes the risk that dissatisfied Muslims might start trying to redraw the map by force. The new leadership in the Serb entity opens up many possibilities of furthering the process of reconciliation, reconstruction and integration, and gives the donor community the excuse it needs to disburse aid more equitably to each entity.

Generous international support for Dodik could also encourage the emergence of new non-nationalistic leaders in the Federation and throughout the region. There is certainly room for improvement in the Federation. In many ways the increasingly nationalistic Bosniak Party for Democratic Action, headed by Izetbegovic, had too free a ride in 1996–97. That the international community leaned too far in support of the Muslim community was based on justifiable sympathy for what it had suffered during the war and the consistently appalling behaviour of the leadership in the Serb entity not only during, but also since, the war. With more responsible leadership in Banja Luka, standards for compliance with the Dayton Accords can also be set much higher for the Federation.

◆

Turkey: If Neither West Nor East, Then What?

There is an old Turkish saying that Turkey is a man running West on a ship heading East. But, by early 1998, these two apparently contradictory currents that had characterised Turkish politics over the previous decade both appeared checked. In June 1996, the Islamist *Refah* (Welfare) Party, under its veteran leader Necmettin Erbakan, had taken power for the first time. Yet the grass-roots Islamist movement quickly received two blows: first, being toppled from power by Turkey's staunchly secularist military; and, second, being banned by the Constitutional Court. At the same time, Turkey's Western-oriented establishment had to come to terms with the announcement by the EU at the Luxembourg summit on 12–13 December 1997 that talks on Turkey's possible accession were to be indefinitely postponed. Nor was Turkey able to find solace elsewhere in the international arena. The former Soviet republics of Central Asia, once touted as offering Ankara a belt of influence stretching from the Caspian to the Chinese border, had not rushed to accept its embrace; while other Muslim states had issued a series of strongly worded condemnations of Turkey's increasingly close ties with Israel.

The resulting bout of national introspection involved more fundamental issues than mere international isolation. Suddenly, the academic question of Turkey's identity seemed to take on a hard and practical edge, involving both strategic alignments and economic interests. For many Turks, 1997 brought the uncomfortable realisation that the division between East and West, if it existed, lay not along the country's eastern or western border, but ran like a fault-line through the heart of their society, leading them to question not only what Turkey's role in the world might be, but how they should see themselves. Both Turkey's Western-oriented élite and its grass-roots Islamists had, to a large extent, defined themselves through their relations with others. Turkey's rejection by the EU raised the question of whether Europe, despite its frequent reassurances, saw Muslim Turkey as truly European and, if not, whether Turks could consider themselves as such. The Islamists, who had maintained that Turkish identity was defined through religion, were having to come to terms with the fact that not only had they been toppled from power by fellow Turks rather than by outside forces, but also that the strongest international condemnation of the banning of *Refah* came not from other Muslim states but from the EU.

In 1997, confusion over Turkey's identity was compounded by a growing disillusionment with the political system amongst the general public. During the course of the year, evidence steadily mounted that special anti-guerrilla units, formed to combat the separatist Kurdistan Workers' Party (PKK) in the long-running insurgency in the south-east of the country, had combined with politicians, local clan chiefs and organised crime to form networks which became involved in narcotics smuggling, underworld assassinations and the creation of death squads to kill suspected PKK supporters. In early 1998, several members of parliament, including a former interior minister, were facing charges of connivance or active involvement in the networks. Further investigations into allegations of bribery and embezzlement were pending against politicians from across the political spectrum.

There was also a widespread acceptance that the existing political system in Turkey had failed to produce solutions to the country's many economic and social problems, particularly persistent high inflation, which has remained at an annual rate of over 60% for more than a decade. As further evidence of this failure, both secularists and Islamists could cite the clash between the *Refah*-led government and the military: the secularists because the political system had failed to prevent the growth in the Islamist movement, and the Islamists because it had failed to prevent the military from intervening in the political process.

The Military Returns

Seventeen years after it last staged a coup, the Turkish military in 1997 again demonstrated that it remains the supreme arbiter of political power in Turkey. Although traditionally a reluctant interventionist, the military had been watching developments with alarm. Both the structure it had left in place when it returned the country to civilian administration in 1983 and a succession of inept and corrupt governments had failed to address the social and economic factors that fuelled the rapid growth in the Islamist movement during the late 1980s and early 1990s. *Refah* had nearly tripled its share of the vote from 8.7% in 1987 to 21.4% in the general elections of December 1995, when it emerged as the largest party with 158 seats in the 550-seat unicameral parliament. In June 1996, it had come to power at the head of a coalition with the centre-right True Path Party (DYP) of former Prime Minister Tansu Çiller.

To a large extent, the rise of the Islamist movement was the result of the failure of the Turkish political system and the failure to overcome economic stagnation. Despite over 50 years of theoretical multiparty democracy, the Turkish political culture remains authoritarian and hierarchical rather than participatory. Turkish political parties still tend to form around a strong charismatic individual, and are held together by personal loyalty and patronage rather than ideological conviction. These factors have encouraged nepotism and corruption, made inter-party cooperation almost impossible and fragmented the vote of both the centre right and centre left between several parties with almost identical manifestos. The result has been a succession of weak coalition governments which have opted for short-term populism rather than radical and potentially unpopular measures to tackle the country's economic, social and political problems.

During the 1990s, *Refah* was the only political party with any ideological momentum, promising a more equitable society based on Islamic values and a reorientation of Turkish foreign policy away from the West to focus on closer ties with other Muslim countries. *Refah* attracted a broad range of support, from pious conservatives to hardline advocates of *sharia* law. But its appeal was strongest amongst the poorer sections of society, who remained most closely wedded to traditional Islamic values and who had seen their earnings eroded by years of high inflation.

By the beginning of 1997, however, after six months in power, *Refah* had introduced no major policy initiatives, nor provided any programme to alleviate economic problems, preferring to bide its time and build up strength by infiltrating party supporters into the civil service. Yet, concern about what it might do made its presence in government anathema to the military, which regards itself as the guardian of the principle of secularism enshrined in the Turkish constitution.

Since Turkey's return to civilian rule in 1983, the military's role in policy making had been gradually restricted to a *de facto* control over security issues, such as the conduct of the war against the PKK. By the beginning of 1997, the military's *de jure* political role was limited to the presence of five members of the military, alongside five members of the civilian government, on the ten-person National Security Council (NSC), the country's supreme advisory body. Despite this, the military continues to enjoy considerable informal influence. Its record of three coups in the last 40 years means that, when it expresses an opinion, its voice is difficult to ignore.

As electoral support for *Refah* rose, the military issued increasingly blunt statements about the inviolability of the secular state. When *Refah* took office, the military moved to implement a two-stage strategy: to remove *Refah* from government, and then to address the social and economic issues which had fuelled its rise to power. The military stressed its commitment to achieving its goals within the system, without recourse to a direct coup, although it also made clear that it would use force if all other methods failed.

In the mid-1990s, the planning departments in the Turkish General Staff had been expanded both to monitor the development of the Islamist movement and to provide an infrastructure of expertise on every aspect of civilian policy making. On 28 February 1997, the military, at a meeting of the NSC, presented the *Refah*–DYP coalition with an 18-point list of policy recommendations. These focused on the extension of compulsory minimum education from five to eight years and stringent restrictions on religious schools and private Koran courses, which the military believed were being used to inculcate anti-secularist values.

Refah was already coming under opposing pressure from more radical elements within the party over its failure to implement its election manifesto. It could afford neither to risk alienating its voters nor to challenge the military by refusing to carry out its recommendations. As *Refah* equivocated, the military began to increase the pressure, holding a series of briefings for the press and judiciary on the dangers of Islamic fundamentalism, while working behind the scenes to persuade members of the DYP to withdraw their support for the government. On 21 May 1997, the Public Prosecutor applied to the Constitutional Court for *Refah*'s closure on the grounds that it had attempted to incite the violation of the constitution's provisions on secularism. Significantly, almost all of the specific charges presented to the Constitutional Court, mostly references to speeches made by *Refah* leaders, referred to incidents that had taken place before *Refah* came to power. On 18 June, his parliamentary majority steadily eroded by defections from the DYP, Erbakan resigned as Prime Minister. On 16 January 1998, the Constitutional Court announced *Refah*'s closure. Erbakan and five other *Refah* deputies were dismissed from parliament and banned for five years from leading a political party.

The military's role in toppling the *Refah* government and the subsequent court closure of the largest party in parliament have again raised questions about Turkey's commitment to the democratic process. Yet the moves were welcomed by the majority of the population, many of whom remain prepared to tolerate limitations on democracy in order to defend what they regard as more important principles, such as secularism and territorial integrity. There are also questions about *Refah*'s commitment to the democratic process. In campaign speeches prior to the 1995 elections, *Refah* leaders had promised once in power to introduce a system 'better than democracy', which increased concern amongst the military that, if it gained power, *Refah* would never give it up.

In early 1998, the medium- and long-term impact of the closure of the WP on the Islamist movement remained unclear. The Islamists appeared likely to regroup in a new party, although there was little doubt that it would take time to rebuild what had been the most efficient political party structure in Turkey. What was less clear was who the leader would be. It was expected that, initially, Erbakan would control any new party by appointing a surrogate as leader. But, at 71 years of age, Erbakan is nearing the end of his political career. The closure of the WP is likely to accelerate the process by which the leadership of the Islamist movement is handed over to a younger, more radical generation. After a transitional period during which the new party is effectively controlled by Erbakan, Turkey's secular establishment could find itself faced with a rejuvenated movement headed by a more hardline leadership; although in early 1998 the vast majority of the younger generation of Islamists remained committed to taking power by peaceful means rather than resorting to violence.

Plus Ça Change

The widespread public relief that the *Refah*–DYP government had been removed from power without a coup gave the new coalition appointed on 30 June 1997 an unprecedented opportunity to introduce a much-needed reform programme. The military indicated that it was willing to allow the new administration a period of grace, and gave it a clear mandate to address the economic and social problems which had nurtured the rise of the Islamist movement. But the new government quickly became bedevilled by the same weaknesses that had marred other administrations prior to the *Refah*–DYP coalition.

The new coalition comprised the centre-right Motherland Party (*Anap*), headed by Mesut Yilmaz, who became Prime Minister, the nationalist-left Democratic Left Party (DSP) and the centre-right Democratic Turkey Party (DTP), which is mostly composed of defectors from the DYP. To achieve a parliamentary majority, however, the coalition needs the support of the social-democratic Republican People's Party (RPP). In September 1997, it introduced eight-year compulsory education. But it appeared unable to coordinate, much less implement, economic policy initiatives as both the coalition partners and members of the same party began to fight over spheres of political responsibility and to manoeuvre for advantage in the run-up to what might be an early election.

On taking office, the coalition had promised to introduce tax reform, accelerate privatisation and reduce annual consumer inflation from 78% at the end of June 1997 to 75% by the end of the year and 50% in 1998. Turkey's persistently high inflation continues to be due to high budget deficits, which are fuelled by loss-making state enterprises, an over-staffed bureaucracy and a grossly inefficient and inequitable tax system, which throws most of the tax burden onto low-paid wage earners. Current estimates put the size of the country's thriving black market at the equivalent of anything from a third to one-half of total gross national product (GNP).

On 23 July 1997, Gunes Taner, one of three ministers appointed by Yilmaz to oversee the economy, declared that inflation would top 100% by the end of the year. In one week in November, each of the three ministers announced a different economic stabilisation package, none of which had been implemented by early 1998. On 1 December 1997, Yilmaz declared a six-month freeze on public sector prices, only to withdraw it the next day. On 5 December, Treasury Under-secretary Mahfi Egilmez, the top economic bureaucrat, resigned in despair. By the end of 1997 annual inflation had risen to 99.1%. In early 1998, amid speculation that early elections would be called later in the year, the government appeared to have neither the time nor the political will to introduce a reform package.

A Bloody Stalemate

There were no fresh initiatives in 1997 to end the 14-year conflict with the PKK in the predominantly Kurdish south-east of Turkey. The insurgency continues to exact a high death toll and to account for the majority of human rights abuses in the country. On taking office, Yilmaz had promised to improve Turkey's human rights record, but doubts remain whether, even if the coalition wanted to, it would be able to change the situation. The south-east remains virtually under military administration. Central government authority rarely penetrates very deeply into the security apparatus, especially to such elements as the specialised anti-guerrilla units known as 'Special Teams'. Although they have been implicated in many human rights abuses, the military command has traditionally allowed them considerable autonomy provided that they continue to inflict casualties on the PKK.

During 1997, there were indications that the scorched-earth policy implemented since the mid-1990s, which has included the forced evacuation of over 2,000 villages and the displacement of more than 1.5 million people, continues to restrict the PKK's freedom of movement. In 1997 most of the fighting was confined to mountainous areas. During the year the Turkish military claimed to have killed over 6,000 militants for the loss of 550 members of the security forces. The accuracy of these figures is open to doubt, and this raises questions over an earlier claim by the military at the end of 1996 that it had killed 2,516 PKK militants during the year and destroyed its effectiveness as a fighting force.

Many of the alleged PKK casualties came during cross-border operations against PKK bases in the Kurdish enclave in northern Iraq. Since 1996, the PKK has been in alliance with the Patriotic Union of Kurdistan (PUK) of Jalal Talabani, which controls the east of the enclave, while Turkey has enlisted the assistance of the Kurdistan Democratic Party (KDP) of Talabani's rival Masoud Barzani, supplying him with arms in return for intelligence and occasional ground support in operations against the PKK. In May 1997, the Turkish Army launched a six-week ground and air cross-border operation involving an estimated 20,000 troops against PKK positions in northern Iraq. Privately, Turkish generals admit that such operations are primarily designed to inflict logistical damage as the PKK's light and mobile fighting units are able to disperse before the troops arrive, while the mountainous terrain provides ample cover against air strikes. In November 1997, the Turkish Army launched another, more limited, cross-border operation targeting several of the same PKK bases, which had been restocked during the summer. Despite their alliance, the Turkish military remains wary of the KDP, noting that PKK militants still cross KDP-controlled territory on their way from the organisation's training camps in the Bekaa Valley via Syria either to Turkey, or to PUK-held areas in northern Iraq. By the end of the year, the Turkish army had established a small permanent presence in northern Iraq, mostly for patrolling and intelligence gathering, and as forward observers for air strikes.

The conflict in the south-east continues to be largely ignored by the rest of the country, despite the high death-toll. Although there is continued official rhetoric about the indivisibility of the Turkish state, no effort has been made either to invest in what is the most impoverished region of the country or to demilitarise the local administration. For most people in the region, their only contact with the state is through often hostile and suspicious soldiers. Not surprisingly, despite the absence of violence in the major cities in the region, PKK support groups and recruiting networks remain very strong, particularly amongst the urban poor, whose ranks have been swelled by the inhabitants of the forcibly evacuated villages.

The End Of An Unconsummated Union?

In the late 1980s, Turkish politicians were fond of describing Turkey as a Muslim country with a European vocation. Many hoped that, within a generation, Turkey would be an EU member as the culmination of a process of Westernisation which began with the foundation of the Turkish republic in 1923. For most Turks, EU membership meant more than mere economic or political benefits. It was a question of being able to number themselves amongst what they perceived as the élite of nations.

The indefinite exclusion of Turkey from the list of candidate countries for imminent accession announced at the Luxembourg summit came as a bitter blow. For most Turks, the sense of humiliation and rejection, which many still suspect was primarily a product of racial and religious prejudice, was compounded by anger at the reasons cited by the EU for judging it ineligible for membership, particularly criticism of Turkey's human rights record and treatment of its Kurdish minority. On 14 December 1997, Prime Minister Yilmaz announced that Ankara would sever all official contacts with the EU, except for the customs union, and restrict itself to bilateral relations with the member-states.

Relations between Turkey and the EU had been strained during 1997 by the failure to resolve the Cyprus problem against a backdrop of increasing militarisation of the divided island. Tensions rose in early January 1997 when the Turkish military responded to the Greek Cypriot purchase of S-300 surface-to-air missiles from the Russian Federation by threatening to use force if necessary to prevent their deployment, planned for late in 1998.

The missile issue prompted a series of agreements between Ankara and the Turkish Cypriot administration to integrate the north of the island into mainland Turkey. To a large extent, *de facto* integration had already taken place. Turkish Cypriots already use the Turkish Lira and Turkish school textbooks, and they can travel to the mainland without a passport. By 1997, an estimated one-third of the 170,000 population of the north had been born in Turkey, while a paucity of job opportunities means that most young Turkish Cypriots migrate to the mainland to pursue a career.

The process of integration accelerated following the July decision by the European Commission to recommend that the EU begin accession negotiations with the Greek Cypriot government in the name of the whole island, and the failure of UN-sponsored face-to-face talks between Cypriot President Glafkos Clerides and Turkish Cypriot leader Rauf Denktash at Troutbeck, New York, on 9–12 July and Glion, Switzerland, on 11–16 August. During 1997, agreements were signed between Ankara and the Turkish Cypriot administration to allow the latter to circumvent its lack of international recognition by using Turkish foreign missions abroad, integrate the Turkish and Turkish Cypriot social security and health systems and make Turkish state aid available to Turkish Cypriot businesses. On 16 December, in the wake of the Luxembourg summit, Ankara and the Turkish Cypriot administration issued a statement declaring that negotiations on the Cyprus problem would be frozen until the north of the island received international recognition as an independent state. In early 1998, such recognition appeared increasingly unlikely and the process of integration irreversible.

A Strategic Realignment In The Eastern Mediterranean?

Turkey's growing estrangement from the EU in 1997 was partially offset by increasingly close ties with Israel. The main driving force was the Turkish military, which regarded Israel as a more pragmatic and less critical ally than the EU, whose member-states had frequently suspended arms sales to Turkey amidst allegations that they were being used for internal repression of the Kurds. The two countries also shared a common enemy, Syria. High-level military and Defence Ministry visits between the two countries continued throughout 1997 and early 1998, building on the 1996 military-cooperation agreement which allowed Israeli aircraft to train in Turkish airspace. In 1997, Israeli firms continued to win lucrative Turkish defence tenders, including a $75m contract to upgrade 48 of the country's F-5 fighter aircraft. This followed a $632m contract in 1996 to upgrade 54 Turkish F-4s. In early 1998, the Turkish military was also reported to be negotiating the purchase of ten unmanned aerial vehicles from Israel, while Israeli companies were lobbying to upgrade Turkey's M-60 tanks and jointly produce *Arrow* and *Popeye* missiles. Relations were further underpinned by the participation, in early January 1998, of Turkish warships in naval exercises, codenamed *Reliant Mermaid*, in the eastern Mediterranean with the US and Israel.

Ankara's *rapprochement* with Israel has inevitably generated considerable suspicion amongst Arab states. The resolutions passed at the Organisation of Islamic Conference meeting in Tehran on 9–11 December 1997 demonstrated that Muslim states were far from convinced by Ankara's insistence that closer ties with Israel would not affect its relations with other states in the region. But, for the Turkish military, Israel remains a natural ally, both because of a shared distrust of Arab states and to counterbalance what it claims is a threat from an anti-Turkish strategic alliance between Greece and Syria.

Significantly, in 1997 Israel was the only country with which Turkey's relations can be said to have improved. In late 1997, there even appeared to be a cooling in relations between Ankara and Washington, which has traditionally been more tolerant than the EU of Turkey's human rights record and its cross-border operations into northern Iraq. In February 1998, Turkey did not wholeheartedly back the US threat to use military force to persuade Iraq to comply with UN resolutions.

Where Now?

In early 1998, almost every section of Turkish society was permeated by pessimism about the future, while disillusion with the political process and the continuation of an economic trough extended across the political spectrum. When they looked ahead, most Turks were unsure not only of Turkey's future place in the world, but also about how to see themselves. Perhaps most damagingly, the circumstances surrounding the toppling of the *Refah*-led coalition had revealed a social polarisation that questioned one of the most cherished of all Turkish sayings, namely that a Turk's only true friend is another Turk.

The withdrawal of the prospect of early EU membership removed one of Turkey's main sources of momentum, whether in harmonising its legislative and regulatory environment with that of the EU, or providing Ankara with an incentive for greater democratisation and an improvement in human rights. The Luxembourg summit, and Ankara's reaction to it, stripped both Turkish foreign relations and much of its domestic policy of a cohesive sense of direction, leaving Turkey facing a prolonged period of injured, uneasy introspection with no clear indication of the timing or nature of the outcome.

Nor has the performance of the coalition between *Anap*, the DSP and the DTP persuaded the Turkish military to reverse its low opinion of the ability of politicians to implement policy even when it has been formulated. Until the military is confident that politics can be left to the politicians, it is likely to stay in the political arena, ready and able to intervene if it considers that the founding principles of the state are at risk. Either directly or indirectly, the military will remain for the foreseeable future the ultimate arbiter of political power in Turkey.

The Middle East

A long year without any progress towards peace left the Israeli government, the Palestinians and neighbouring Arab states further apart in early 1998 than at any time since the signing of the Oslo Accords in 1993. The stalemate, with suspicions rising and hopes of a breakthrough falling on both sides, has become dangerous. The uneasy situation of neither peace nor war may last for some time, but it is doubtful whether violent clashes can be avoided unless the Palestinians are offered more hope. Only the US has the weight and influence to get the talks going again, and, by the end of March 1998, it appeared to be prepared to take action.

The US had also had to take action in Iraq, forcing Saddam Hussein to end his intransigence towards the United Nation's Special Commission inspectors. Faced with a massive build up of US military power in the Gulf and an escalating drumbeat of threats to use it to better effect than previously, Saddam agreed to proposals brought by UN Secretary-General Kofi Annan which would allow further inspections. But he had not lost his gamble entirely. He had widened the split in the Security Council over how to deal with him and had made the day when sanctions would be lifted move much closer.

Other changes were taking place in the Gulf. The election of a moderate cleric, Mohammad Khatami, as president of Iran in May 1997 transformed domestic politics. His advent has also had a significant impact on foreign affairs. Iran has made extensive efforts to improve relations with Saudi Arabia, and has even sent out diplomatic feelers to Iraq and Syria. Perhaps most dramatically, Khatami has made it clear that he would like improved relations with the US. It is still uncertain whether Khatami would win if it came to an open struggle with the still powerful hard-liners, but, for the first time since the revolution, there is hope that key national policies may be moderated.

◆

No War, No Peace

While most observers agree that a negotiated settlement between Israel and its neighbours remains the only way to break the vicious circle of violence that has afflicted the Middle East for decades, very few think there will be one soon. The air of optimism generated by the famous Rabin–Arafat handshake on the lawn of the US White House in September 1993 dissipated long ago. The sense of futility on both sides has not been so clearly expressed since the Israeli–Palestinian peace process began in Oslo in 1993. Israeli official statements refer to the process as going through a critical stage. The Secretary-General of the Palestinian Cabinet, Ahmed Abdel Rahman, has gone much further. In March 1998, he announced the death of the peace process despite the absence of an official memorial service.

Some analysts argue that a regional stalemate is hardly a novel phenomenon in the Middle East. While this is true, the current stalemate is particularly dangerous because it must be measured against the region's increased expectations that the threat of war was lifting and peace was arriving. In addition, it is particularly dangerous because it appears to be untenable. It cannot persist for long without creating an atmosphere conducive to acts of violence and terrorism that will further harden the attitudes of Israel, the Palestinians, Syria and Lebanon. Moreover, relations between Israel and its Egyptian and Jordanian neighbours, who have signed peace treaties with the Jewish State, cannot be insulated against the severe repercussions of such regional stalemate and deterioration. The failure of concerted efforts at a normalisation of regional economic relations between Israel and the Arab states highlights this problem. Thus, while outright war in the region remains unlikely, the inability of the parties in the dispute to nudge the process forward could lead to a new round of violence.

Israel's Ruling Coalition

Binyamin Netanyahu came to power in May 1996 with a programme to modify the Oslo Accord, not to cancel it. At the core of that 'peace with security' programme were two concepts, namely the linkage between achieving Palestinian autonomy and maintaining Israeli security, as well as the notion of reciprocity between the fulfilment of obligations by the two sides. Netanyahu's election represented a watershed in Israeli politics: it was the first time that the people had directly elected a prime minister. Previously, it was the task of the Israeli president to decide who should be asked to form the government, given the election results. Although Netanyahu won by a slim margin, he was able to claim a political mandate independent of the outcome of Israeli parliamentary elections.

The succeeding events of 1997 made it clear that Israel's redeployment from most of Hebron as a result of the agreement signed on 15 January was not an example that would be followed in further implementation of the Oslo Accord. It was premature for an enthusiastic Arafat to tell his Palestinian supporters on 20 January: 'We have concluded a peace agreement with the entire Israeli people... The 87 Knesset votes in favour of the agreement represent a new reality in the Middle East'.

Under pressure from his partners in the coalition government, Netanyahu has since then insisted that his government alone must decide any future Israeli deployment. Ambassador Dore Gold, the Israeli representative to the United Nations and a Netanyahu confidant, articulated that position in July 1997 when he stated that 'further redeployments are issues for Israel'. Neither negotiation with the Palestinians nor pressures by a third party – namely the US – have moved Israel from the belief that it alone can determine the extent of further Israeli deployments (see map, page 232). More than half of the West Bank has been designated by the Netanyahu cabinet as necessary for Israel's security.

In addition, the support of Jewish settlements and the right of Jews to settle in the 'greater land of Israel' loom large in Netanyahu's statements and in the political platforms of the parties that form the ruling coalition in Israel. According to the Israeli newspaper *Ha'aretz*, the Jewish settler population of the West Bank grew by 7.45% during the first ten months of 1997. This was the equivalent of a 9% growth rate for the year, only slightly down from the 9.4% increase in 1996. According to Israeli statistics, the Jewish population in the West Bank has reached over 155,000. The largest growth came in areas near sizeable Palestinian population centres in Nablus, Ramallah and Bethlehem.

While encouraging more Jewish settlements in the West Bank is in line with Netanyahu's own ideological beliefs, it was also useful as an instrument to cement his ties with his partners on the right of the ruling political coalition. Similar ideological issues have exercised significant influence over the government coalition-building and coalition-maintenance processes. The natural government coalition constituency for Netanyahu following the 1996 elections was with the religious and right-wing nationalist parties. The main winners in the Knesset elections were smaller parties, especially those with a religious base.

Given the narrow margin of the ruling coalition, the government needed the support of the small parties, which were able as a result to exercise disproportionate influence over government policy. This was true, for instance, with regard to the National Religious Party, United Torah Judaism, *Shas* and *Tsomet*, all parties which advocated an activist settlement policy. *Likud* hawks such as former Defence Minister Ariel Sharon also supported that policy. The Netanyahu government's plans to double the number of Jewish settlers in the Occupied Territories by building or expanding Jewish settlements was intended to create a sense of irreversibility of the settlement projects. Even as Israeli–Palestinian peace talks were taking place in Washington in November 1997, Israel's representatives refused to stop building and expanding settlements. The strong influence of settlers on the *Likud* and most of its coalition partners goes far to explain that position. Incentive programmes that were reinstated by Netanyahu's government encourage still more people to move to the West Bank settlements.

Netanyahu's policy on settlements has served to assure religious and ultra-nationalist parties that he was not the kind of political leader who comes to power on the basis of a tough political programme, only to make major concessions later. However, it confirmed the worst expectations of the Palestinians and the Arab leaders that Israel's policy under Netanyahu would renege on the 'land for peace' formula adopted by former Prime Minister Yitzhak Rabin and his successor, Shimon Peres.

Arafat's Predicament

As part of his belief in reciprocity, Netanyahu has insisted that the Palestinian Authority (PA) is obliged to stamp out terrorism on its own territory or emanating from there to Israel. This has put Yasser Arafat in an difficult position. Whenever there is a new attack, Netanyahu and his government accuse Arafat of being responsible for the violence and insist that the PA curb, if not eliminate, the infrastructure of violence in areas under its control. Israel's next step has often been to close the border to Palestinian workers, freeze any further action on the peace process and build more Jewish settlements. A vicious cycle is thus established. Under these circumstances, it is difficult for Arafat and the PA to show the Palestinian people any real fruits of peace, and the terrorist organisation *Hamas* gains both credibility for its position and personnel for its further actions. Arafat is thus caught between Israel and *Hamas*, while having very little power to influence their positions or actions.

Threats of violence against Israel by *Hamas* can be triggered by a host of developments over which Arafat has little control. On 30 July 1997, two *Hamas* suicide bombers killed 16 people in Jerusalem. *Hamas* was behind another incident in a pedestrian shopping street in Jerusalem in September 1997, in which five Israelis died. Efforts to resume the talks were effectively torpedoed.

In mid-February 1998, the armed wing of *Hamas*, *Izzadin-el-Qassam*, threatened to strike at Israeli civilians if the US decided to launch a military attack against Iraq. As Israelis were worried about the likelihood of an Iraqi missile attack and violent *Hamas* acts, Arafat had to accept the fact that most Palestinians under PA control were sympathetic towards Iraqi President Saddam Hussein. Demonstrations called on Saddam Hussein to hit Tel Aviv with missiles and to use chemical weapons against Israel. Arafat explicitly banned these pro-Iraqi demonstrations, putting himself at odds with most Palestinians. His main objective was to avoid antagonising Washington, since in his view, only the US could get the peace process moving again. At the same time, Arafat has been reluctant to antagonise *Hamas* and to provoke its leaders into a confrontation with the PA. It was for this reason that he had decided to include *Hamas* in the National Unity Conference that was held on 20–21 August 1997.

Although they have made no progress, the Palestinians have continued to conduct high-level negotiations with the Israeli government whenever possible. Arafat himself has only met Netanyahu once, on 8 October 1997. He made other meetings with the Israeli Prime Minister contingent on concrete progress in the peace process. His top aides, Mahmoud Abbas (Abu Mazen), Saeb Erikat, Nabil Sheath and the security chief Mohamed Dahlan met Netanyahu's top assistants many times during 1997. Topics discussed in these meetings included the features of the transition period, the conditions of building a Palestinian airport and seaport in the Gaza Strip, and the corridor linking Gaza and the West Bank, as well as final-status issues. Some of the talks took place at the residence of the US Ambassador to Israel, reflecting an attempt to make the US a witness to the negotiations. Arafat also agreed to allow the US Central Intelligence Agency (CIA) participation in security talks between the PA and Israel in August and November 1997.

None of these talks has suggested that any progress on the many issues discussed was remotely possible. In his speech before the Palestinian legislative council on 7 March 1998, Arafat warned that the peace process was 'about to breathe its last breath'. Some of the rhetoric is intended to deepen concern in the hope of inducing new action, but with no progress to show in more than a year, Arafat's statement is uncomfortably close to the truth. Arafat's political destiny is linked closely to the future of the peace process. Nor is his position made any stronger by reports that he is in ill health. One positive factor in his favour is the difficulty of finding a successor.

Yet Arafat has also to deal with the restless members of the legislative council who have demanded a larger role for the Palestinian public in the Authority, and more accountability from the PA's current members. Arafat's ability to deal with political institutions has been openly criticised and there have been threats of a vote of no confidence in his government. Members of the legislative council have expressed both dismay and anger over Arafat's refusal to put an end to corruption, mismanagement and waste in the PA and to sign the laws passed by the 88-member legislature. His empty promises of reform and his emotional appeals for national unity to appease the legislators have lost most of the little credibility they might once have had. In July 1997, after uncovering large-scale corruption among high-ranking PA officials, they voted 56 to 4 in favour of a resolution calling on Arafat to dissolve the cabinet by September 1997. Nor is the criticism confined to the legislative council. According to a subsequent public opinion poll conducted by the Jerusalem Media and Communications Centre, about 63% of Palestinians believe that corruption is rampant in Arafat's PA.

With persistently high Palestinian unemployment and deteriorating living standards, more settlements being built or expanded and the peace process at a standstill, Arafat's position has grown weaker daily. At the end of 1997, he began warning about the dangers of a new Palestinian *intifada*. He received unexpected support when an open letter was sent to Netanyahu in early March 1998 by 1,154 Israeli army officers, including a former chief of staff, 11 generals, and 216 officers of the rank of lieutenant-colonel or higher. The letter warned that there might be more conflict and another *intifada* as a 'historic window of opportunity is about to slam shut'.

Instances of violence pitting Palestinian youths with rocks against Israeli troops with rubber bullets have increased. Another *intifada* may be in prospect, but whether it would be led by Arafat is difficult to predict. His position is so badly jeopardised by the current stalemate in the peace process and the sense of despair and rage among the Palestinians that he may feel he must assert his leadership again in this way. If he does, the peace process will probably suffer a mortal blow.

Ready To Leave South Lebanon?

Since the beginning of 1998, Israeli leaders have turned their attention away from dealing with the stalemate in Israeli–Palestinian peace negotiations and focused instead on Lebanon. They signalled their readiness to withdraw Israeli troops from the security zone Israel established in south Lebanon in June 1985 if the Lebanese government provided Israel with adequate security guarantees. Although the basic concept is agreed, the content of the guarantees Israel would require remains unclear. Such guarantees were reportedly not agreed by the top political and military leaders themselves. It is not clear whether the Lebanese Army's prevention of *Hizbollah* attacks against northern Israel would be enough, or if Israel would demand the disarming of *Hizbollah* to enable the Lebanese state to extend its military and political control over the approximately 10% of Lebanon now under Israeli control.

What is clear is that, for the first time, Israel has dropped its customary demand that the conclusion of a peace treaty with Lebanon would be a necessary condition for Israeli troop withdrawal from the country. Moreover, three new developments may prove to have significance with regard to such a proposal. The first emerged during Netanyahu and Defence Minister Yitzhak Mordechai's trips to Europe in February and March 1998, when Israel accepted a vague linkage between a security deal with Lebanon and resuming its peace talks with Syria. Second, Israeli leaders sought to enlist European assistance in providing guarantees that

Figure 6 UN Security Council Resolution 425 (19 March 1978)

The Security Council,

Taking note of the letters of the Permanent Representative of Lebanon (S/12600 and S/12606) and the Permanent Representative of Israel (S/12607),

Having heard the statements of the Permanent Representatives of Lebanon and Israel,

Gravely concerned at the deterioration of the situation in the Middle East, and its consequences to the maintenance of international peace,

Convinced that the present situation impedes the achievement of a just peace in the Middle East,

1. **Calls** for strict respect for the territorial integrity, sovereignty and political independence of Lebanon within its internationally recognized boundaries;
2. **Calls** upon Israel immediately to cease its military action against Lebanese territorial integrity and withdraw forthwith its forces from all Lebanese territory;
3. **Decides**, in the light of the request of the Government of Lebanon, to establish immediately under its authority a United Nations interim force for southern Lebanon for the purpose of confirming the withdrawal of israeli forces, restoring international peace and security and assisting the Government of Lebanon in ensuring the return of its effective authority in the area, the force to be composed of personnel drawn from states members of the United Nations;
4. **Requests** the Secretary-General to report to the Council within twenty-four hours on the implementation of this resolution.

would permit Israel to withdraw from south Lebanon without much risk. They seemed to imply that the issue of Lebanon could prove to be an opportunity for Europe to enhance its role in the Middle East peace process. Israel tried to obtain political help, particularly from France, which maintains close relations with Lebanon and Syria. Third, even some Israeli hawks have supported a unilateral, gradual pull-out of Israeli troops if it is linked to security guarantees or to a warning of an overwhelming Israeli retaliation if there are attacks over the border after the redeployment.

This Israeli diplomatic offensive comes against the background of the continuation, even the escalation, of the war of attrition launched by *Hizbollah*, with Syrian and Iranian support, against Israel. This war has turned out to be costly both in terms of lives lost and financial resources, as well as destabilising Israeli communities in the north. At least 39 Israelis were reportedly killed in 1997, the highest annual loss since the security zone in south Lebanon was established. More devastating was the loss on 4 February 1998 of 73 Israeli soldiers in a mid-air helicopter crash on their way to Lebanon. With more casualties and no likelihood of ending *Hizbollah*'s attacks, pressures by the Israeli public for a policy change have mounted. While any major change in Israel's policy toward the Palestinians could have serious repercussions for the ruling coalition's stability, the same does not apply to Israel's increasingly burdensome self-declared security zone in south Lebanon.

The Lebanese Army could control *Hizbollah*, as it disarmed the other militias in Lebanon, if it were assured of Syrian support. There was no sign by mid-March 1998, however, that Syria believed that it would be in its interest to make an Israeli exit from south Lebanon easy without a parallel Israeli withdrawal from the Golan Heights. Syrian newspapers, which reflect the views of the top leadership, rejected Israel's initiative, calling it 'a trap' and a 'deception'. The Assistant Secretary-General of the ruling *Ba'ath* party Abdallah al-Ahmar described it as an 'old ploy'. According to Syria, implementing UN Security Council Resolution 425, which unequivocally calls for the withdrawal of Israeli forces from Lebanese territory, does not require any negotiations, terms, arrangements, restrictions or conditions.

While Israel strives to separate the Syrian and Lebanese fronts, Syria's President Hafiz al-Assad is keen on linking them in order to pressure Israel via a surrogate battlefield in south Lebanon. This situation will continue as long as Israel and Syria fail to agree on the conditions for resuming the talks between them that were suspended in early February 1996. Moreover, if Syria's leaders are convinced that Israel's withdrawal proposal is either a political ploy or an unavoidable policy change, their motivation to cooperate could be considerably reduced. Given Syria's influence in Lebanon, it was not surprising that the Lebanese government rejected Israel's initiative and insisted on unconditional withdrawal. Thus, without Syria's explicit or implicit endorsement, the Israeli initiative will have few chances of success and the Lebanese problem may haunt Israel for years to come.

Frozen Normalisation

The lack of movement in normalising relations between Arabs and Israelis reflects the political impasse in the peace process. Most of the Arab countries in the Gulf region and in North Africa have put their trade relations with Israel on hold; economic interactions have been reduced to a trickle. Opinion polls in these countries suggest that the public strongly favours linking any normalisation of relations with Israel to progress and achievement in the peace process. A meeting of the Council of Arab Parliaments in Sanaa, Yemen, in mid-March 1998 called for the suspension of all forms of normalisation and the closure of diplomatic offices if Israel did not resume the peace process and accept the principle of 'land for peace'. Egyptian businessmen who were attracted to Israel for economic purposes have stopped trying to visit Israel in the present atmosphere.

Jordan's relations with Israel were strained in September 1997, when two Mossad agents posing as Canadian tourists tried to assassinate Khaled Meshal, the politburo chief of *Hamas*, on Jordanian soil. King Hussein returned the agents to Israel, but only after the promise of release from Israeli jails for as many as 70 Palestinian prisoners, including the spiritual leader of *Hamas*, Sheikh Ahmed Yassin. In addition, Hussein is reported to have suspended intelligence cooperation with Israel until Mossad chief, Danny Yatom, was removed.

Both governments worked hard during the following months to ease the tension which had so badly affected their relations. By March 1998, there had been some improvement. In March, Netanyahu dropped Yatom from his position as head of Mossad, and Israeli Trade Minister Natan Sharansky reached a trade agreement with Jordanian officials to reduce tariffs. He and Jordanian Trade Minister Hani Muliq signed an agreement to establish an Israeli laboratory in Amman to ensure the compliance of Jordanian goods with Israeli specifications. They also signed a letter urging the European Union to exempt joint Jordanian–Israeli products from taxes to increase their access to European markets. Despite this, and a visit by Crown Prince Hassan to Israel in the second week of March 1998, resistance among Jordanians to normalising relations with Israel, including cultural relations, remains strong.

Nothing illustrates the tensions between Israel and most of its Arab neighbours better than the failure of an economic conference held in Doha, Qatar, in late November 1997. The conference was intended as part of a series which would produce the economic foundations for peace in the region. Despite intense US efforts to make this conference a success, most of its allies, including Egypt, which hosted a similar meeting in 1996, Saudi Arabia, the United Arab Emirates and Morocco joined Syria and Lebanon in boycotting the conference to protest against the stalled Middle East peace process. Those few states that attended the conference were represented by low-level delegations. Arab businessmen who came tended to avoid their Israeli counterparts.

The conference's failure demonstrated two things. First, the influence of political considerations over economic interests remains strong in the region, and could not be overcome, despite the efforts of top US officials, including Secretary of State Madeleine Albright. Consequently, this 'non-summit', as it became known, was also the most politicised economic conference in the series thus far. Second, Washington's failure underscores the limits of US ability to determine developments in the region either by influencing Israel to make specific political concessions or by encouraging more of its Arab allies and friends to attend a conference that the US considered very important.

The US Role

With the crisis of confidence between the Arabs and the Israelis mounting, many looked to Washington to help break the stalemate and get the troubled peace process back on track. US attempts to activate the process and to encourage the Palestinians and Israel to pursue fast-track negotiations began with visits to the region during 1997 by Special Envoy Dennis Ross. His talks made little progress as neither Israel nor the PA were ready to make any major decisions before Albright's proposed visit to the region in September 1997. The importance of trying to avert the total demise of the peace process was underscored by two suicide bombings in Jerusalem shortly before Albright's week-long tour. During her talks, she focused on a formula of 'security-for-land', a trade-off between a Palestinian crackdown on the militant groups and an Israeli agreement on a significant and credible plan for implementing the delayed stages of troop withdrawal from the West Bank.

Although neither Netanyahu nor Arafat wanted to antagonise the US, both were concerned about the domestic repercussions of accepting such a trade-off. For Arafat, a military confrontation with *Hamas* to crush its infrastructure would have meant a disastrous Palestinian civil war. For Netanyahu, a large withdrawal from Judea and Samaria could have led to the collapse of his shaky coalition.

The next major step Washington took after Ross and Albright's unproductive visits was an exercise in presidential involvement. Relations between the US administration and the Netanyahu government over issues such as Jewish settlements and delayed troop redeployments from the West Bank had dropped to such a low point that when Netanyahu had visited the US in November 1997 to speak to Jewish groups, President Clinton had avoided meeting him. At the turn of the year, he decided to make a new try. In January 1998, he met separately with Netanyahu and Arafat at the White House. He hoped to encourage them to break the diplomatic stalemate that had lasted for over a year. In response, Netanyahu publicly proclaimed that no 'external pressures' could dictate Israeli security policy. In his view, Washington could propose solutions, but it should not try to impose its will on its Israeli ally. In these circumstances, the President's participation had no more effect than those of his subordinates.

Difficult choices still face US policy-makers. Administration officials have identified a four-point agenda: security; implementing further (but delayed) Israeli troop redeployment; a 'time out' on unilateral actions that can jeopardise peace negotiations (such as building new settlements); and the possibility of moving towards final-status talks between Israel and the

Palestinians soon. But more difficult than merely identifying the components of such an agenda is deciding on the direction of the US role and the blend of enticements and pressures needed to achieve it.

It has been rumoured, for instance, that the US might publicly propose that Israel pull back from 13.1% of the West Bank over a period of three months. The Israeli government would consider any such proposal a US ultimatum which would leave them little choice but to accept it or risk the loss of US support. It has thus been trying desperately to avoid making that choice. In an attempt to pre-empt US action, the Israeli cabinet took the unusual step on 22 March 1998 of voting unanimously to turn down the rumoured proposal before it could be offered. Just before this, Israeli sensitivity to the problem posed by its settlement construction could also be seen in the government's reaction to the March 1998 visit to Har Homa by UK Foreign Minister Robin Cook, in his role as representative of the EU, whose presidency the UK holds until July 1998. Claiming that he would have been better advised to visit the Wailing Wall or the Holocaust memorial, Yad Vashem – rather than the building site that Israelis believe is an essential part of a united Jerusalem – Netanyahu cancelled a final dinner for Cook. No Israeli official turned out to see him off at the airport after the standoff, either.

If the US does present its rumoured plan, the PA is likely to accept it, although some of its leaders may consider it both inadequate and late. On 20 March 1998, however, it was still unclear if the US administration would risk taking this stand and if it could mobilise enough Congressional and domestic support for a tough confrontation with Israel. There are significant differences at the high levels of the administration about the contents of any US plan and the mode of its presentation. Yet for the US to be able to help put the Middle East peace process back on track, it will have to clarify for itself, and for the participants, exactly what its policy is. The situation has reached such a serious point that a decision must be made soon if it is to have any positive effect.

Iraq: An Unresolved Crisis

Once again, Saddam Hussein's unique combination of hubris and calculation threw the world into crisis. During 1997, his frustration with the international sanctions regime became ever clearer, while his grip on domestic power seemed to be stronger than it had been for a number of years. Baghdad had reasserted its leading role in the politics of Iraqi Kurdistan where the Patriotic Union of Kurdistan (PUK) and the Kurdish Democratic Party (KDP) were still locked in conflict and were both in negotiation with the Iraqi government. With the relative decline in the fortunes of Saddam Hussein's eldest son, Uday, and the rise of his second son Qusay, there was a temporary realignment and reconciliation of a number of the clans closest to the heart of the Iraqi regime. Furthermore, the 'oil-for-food' deal of UNSCR (UN Security Council Resolution) 986 promised greatly increased resources for patronage.

Accordingly, Saddam Hussein became increasingly confident that he could get what he wanted from the outside world. When this failed to happen, he provoked a series of crises involving the weapons inspections being undertaken by the United Nations Commission on Disarmament (UNSCOM), believing that this would oblige the UN Security Council to focus on the sanctions regime and the conditions for its removal. This was a high risk strategy since it courted the possibility of military action by the US and some of its allies. It also touched on an area of vital Iraqi security interests and threatened an extension of the sanctions regime.

Nevertheless, these were risks which Saddam Hussein was willing to run. He calculated that military action against Iraq, were it to follow the pattern of previous 'punitive' strikes, would not inflict any serious damage on his regime. On the contrary, it could enhance his standing in Iraq and in the Arab world at large, promoting his image as a heroic victim of superpower aggression. Furthermore, the debate about possible military action against Iraq had a divisive effect on the permanent members of the UN Security Council. This, in turn, was thought by the Iraqi government to be useful in its attempt to counter the apparent determination of the United States and the United Kingdom to keep the sanctions regime in place as long as Saddam Hussein remained in office.

Setting The Scene

Iraqi frustration showed itself initially in criticism of the implementation of UNSCR 986. Saddam Hussein complained in March that Washington was deliberately delaying the imports of food and medicine provided for by the resolution, despite the resumption of Iraqi oil exports in December 1996. In fact, the first shipments of foodstuffs reached Iraq at the end of March, ready to be distributed through the Iraqi ration system, but under the eye of UN monitors who were in place by April. Nevertheless, the Iraqi authorities became increasingly annoyed at the slow rate of approval for import contracts under the scheme. They claimed that these delays were largely vitiating the intended benefits of the arrangements, and they repeatedly blamed this on the malign influence of the US behind the UN monitoring machinery. By June, so aggrieved were the Iraqi authorities that, although the UN agreed to the renewal of the 'oil-for-food' arrangements for another six months, Iraq halted oil exports until a system more to its liking had been worked out, both for the pricing and sale of oil and for the importation of goods.

Iraqi complaints focused at this stage on the workings of the 'oil-for-food' deal, but behind this there was a growing frustration about the larger question of sanctions. Saddam Hussein may have believed that his agreement to the terms of UNSCR 986 would lead to the gradual reduction of sanctions as Iraq was re-integrated into the world market. Instead, Iraq found itself ensnared in a mesh of regulations, controlled and policed by the UN, which considerably weakened any of the strategic benefits that might accompany the renewed sale of oil. At the same time, it appeared that UNSCR 986 was intended to reinforce the sanctions regime as a whole by seeking to neutralise international humanitarian concern about its effects. As the first shipments of goods were arriving under the scheme, it became clear that this was not the beginning of the end of sanctions.

In early March 1997, the Security Council renewed the sanctions for a further two months, on the basis of an unfavourable UNSCOM report concerning, in particular, Iraq's chemical and biological weapons programmes. Later in the month, the US Secretary of State, Madeleine Albright, gave a clear public indication that, as far as the US administration was concerned, the sanctions regime would apply to Iraq as long as Saddam Hussein remained its leader. In April, the UN Commission on Human Rights concluded its session with a report strongly condemning human rights abuses by the Iraqi government and renewed the mandate of the Special Rapporteur on human rights in Iraq for a further year. Taken together, these developments confirmed Saddam Hussein's fear that the US and its allies would ensure that the sole relevant condition for lifting sanctions would be the destruction of his regime. Yet Iraq was aware that not all the members of the Security Council shared this view and it worked to exploit these differences.

It thus became part of Iraqi strategy to focus international attention on the question of disarmament as the sole precondition for the lifting of sanctions. This determination underlay the series of crises provoked by Iraq in 1997–98. In April 1997, the head of UNSCOM, Rolf Ekeus, was complaining about a number of apparently minor incidents which indicated a marked deterioration in Iraq's willingness to cooperate with the weapons inspections. Given the growing unease within UNSCOM and at the UN generally about the extent and viability of Iraq's programme for developing biological weapons, this systematic obstruction began to look increasingly sinister. So concerned had Ekeus become that, just before the end of his term of office in June, he urged the UN to take a clear and strong stand against Iraq's persistent blocking of UNSCOM teams in Iraq. As a result, the UN Security Council passed Resolution 1115 whereby Iraq was given until October to cooperate fully with all UN inspection demands or suffer further sanctions (in the form of a ban on international travel by Iraqi officials deemed responsible for the non-cooperation).

Ekeus handed over responsibility for UNSCOM in July 1997 to the Australian Richard Butler. In doing so, he took the opportunity to list Iraq's deceptions and non-cooperation during the period of his mandate, and the impediments it had placed in his path. He underlined the need for his successor to come to grips, not only with the details of Iraq's still troubling chemical and biological weapons programmes, but also with the elaborate Iraqi deception strategies which had enabled Iraq to preserve a capability for producing weapons of mass destruction even after six years of intrusive inspections and monitoring.

First Act Of Defiance...

Butler soon understood the nature of the challenge. His first report to the UN Security Council in October 1997 gave a disturbing picture of the biological and chemical weapons still thought to be in Iraq's possession and referred to warheads and missiles that were still unaccounted for. At the same time, he was sharply critical of the obstruction encountered by UNSCOM, and urged the UN to take decisive action to oblige Iraq to abandon its systematic deception. This produced UN Security Council Resolution 1134 which renewed the UN sanctions until April 1998 and threatened Iraq with a travel ban on Iraqi officials from that date if it failed to comply fully with UNSCOM's requests for full cooperation and unrestricted access within Iraq.

For the Iraqi government, the cost of defiance was a small one to pay for the visible disarray that ensued among the permanent members of the Security Council. Russia, China and France refused to endorse even the proposed mild extension of the international sanctions regime and abstained when the vote was taken. This development encouraged Saddam to further obstruction. Within a few days, the Iraqi National Assembly called on the President to suspend all cooperation with UNSCOM until a timetable for lifting sanctions had been set by the UN. Seeking to exploit further the apparent rift between the US and some of the other permanent members of the UN Security Council, Iraq's Revolutionary Command Council then declared that Iraq would only cooperate with UNSCOM inspection teams if US citizens were excluded, alleging that the US members were engaged in espionage for their own state, rather than acting on behalf of the UN. It thereupon ordered all US citizens attached to UNSCOM to leave Iraq within a week. By the end of October, the UN had suspended all UNSCOM operations in Iraq and the Iraqi government had set out its conditions for further cooperation. These included setting a timetable for ending inspections and lifting sanctions, ending the US Air Force U-2 high altitude missions and the inclusion of a higher proportion of non-US citizens in the UNSCOM teams.

In the first half of November, the crisis unfolded against a background of threatened military action by the US, attempts by the UN Secretary-General to resolve the issue through mediation and a series of acts of defiance by the Iraqi authorities. In early November, the UN special mission left Baghdad empty-handed and an acrimonious dispute broke out between the Iraqi authorities and the UNSCOM head, Butler. Iraq accused him of breaching understandings on sensitive sites, which required prior notification before inspection, and it simultaneously extended the list of these sites. For his part, Butler accused the Iraqi authorities of removing and tampering with the monitoring equipment installed by UNSCOM at a number of sites throughout Iraq. On 12 November, the Security Council passed Resolution 1137 unanimously. This prepared the way for the travel ban on Iraqi officials which some members of the Council had refused to endorse only a fortnight before. Nevertheless, Iraq maintained its defiance, renewing threats to shoot down U-2 planes (which had resumed their overflights unchallenged, in fact, by Iraqi air defence) and expelling all US members of the UNSCOM teams.

The Iraqi strategy appeared to be yielding dividends. When faced by a direct threat to the UNSCOM weapons inspection team, the US administration reinforced its military presence in the Gulf, apparently in readiness for punitive military action. At the same time, however, Washington hinted that it might be open to a diplomatic compromise in which it might link the ending of sanctions to a satisfactory report on Iraq's weapons of mass destruction, rather than to ending Saddam Hussein's regime. This provided the opportunity for Russia to become more actively involved in seeking a negotiated solution to the crisis. After a visit by Iraq's Foreign Minister, Tariq Aziz to Moscow, the foreign ministers of the US, Russia, France and the UK met in Geneva and agreed to an emergency meeting of the UNSCOM panel to monitor the progress of disarmament in Iraq. For its part, the Iraqi government declared the crisis at an end and announced that the UNSCOM teams, including their US members, would be allowed back into Iraq to continue their work. Although the US and the UK, in particular, were adamant that no concessions had been made to Iraq, it was evident that Saddam Hussein had gained something from the manufactured crisis.

In the first place, both the US and British governments now referred to the possible ending of sanctions when all 'relevant' UN resolutions had been complied with by Iraq. For Saddam Hussein, this indicated that they had accepted that the lifting of the sanctions should be tied only to the UN Security Council resolutions on Iraq's disarmament. There was also some expectation that Russia would act on Iraq's behalf within the Security Council in seeking an end to sanctions. Furthermore, general opposition to American threats of military force, within the region and beyond, could only give encouragement to the Iraqi regime.

Finally, the convening of the UNSCOM panel at the end of November focused UN attention directly on the question of weapons inspections. Given Iraq's record of deception and non-cooperation, this led to an immediate condemnation of Iraq's systematic concealment tactics and renewed demands for unconditional and unrestricted access. However, the panel also seemed to accept the idea that the files might be closed in the near future on certain areas of Iraq's weapons programmes, leading to long-term, less intrusive monitoring systems and holding out the possibility that the UNSCOM mission might soon be ended in its present form. At the November meeting, it appeared that this was the position of the panel on Iraq's nuclear programme and that the same might soon apply in the case of missile development. The chemical and biological weapons programmes, however, remained major items of concern, given Iraq's proven ability to deceive the UNSCOM teams during the previous six years.

Consequently, soon after the UNSCOM teams returned to Iraq, with renewed determination to uncover the details of these programmes, the Iraqi authorities began to raise objections. In particular, they wanted to deny UNSCOM access to certain sensitive and 'presidential' sites which were claimed to be off-limits because they formed part of Iraq's 'sovereignty and national security'. The list of these sites was substantial and some of them were extensive. Denying access to them, therefore, set clear limits on the effectiveness of the UN weapons inspections.

There appear to have been at least two reasons behind Iraq's stance. In the first place, it is almost certain that Saddam Hussein was hoping to retain part of the Iraqi biological and chemical weapons programme. With the talk of a possible 'closing of the files' and a change in the role of UNSCOM, there was evidently some belief that Iraq could emerge with a portion of its arsenal intact. For Saddam Hussein, given his view of the predatory nature of Iraq's environment, retaining some form of deterrence is vital. Equally importantly, he was determined not to let the issue of sanctions, and the link between their removal and ending the weapons inspections disappear from the international agenda. Risky as the strategy might appear, since the two objectives seemed to be contradictory, it was nevertheless important that Iraq should keep up the momentum, especially since the sanctions were in any case going to be maintained until at least April 1998.

And Another...

In these circumstances, Butler's visit to Baghdad in December to try to resolve the growing crisis could only fail, and the Iraqi authorities repeated that presidential sites were off-limits. Despite the demand by the Security Council for full access to suspected weapons sites and the assertion that Iraq was in breach of UN resolutions, the Iraqi authorities extended the categories of 'forbidden' sites to include all presidential palaces, all private property (except with the owner's permission) and the key installations of the Republican Guard and the intelligence services. For its part, the US maintained the forces it had built up in the region, and the US administration hinted once more at the use of force if access by UNSCOM was systematically obstructed.

As if to underline its indifference to the cost of pursuing such a policy, Iraq simultaneously suspended oil exports to finance the 'oil-for-food' deal, arguing that a more satisfactory arrangement for distribution of goods reaching Iraq must be worked out. This was agreed by the UN and in early January a new distribution plan was approved and Iraq resumed pumping oil. Meanwhile, the UN Secretary-General continued work on a report aimed at increasing the amount of oil Iraq was to be allowed to sell under UNSCR 986. Although officially denied, this seemed to be a direct outcome of the November crisis when the UN had been looking for ways to induce Iraq to back down from its position on the work of UNSCOM and the US role therein.

Welcome as the possible increase in revenues might have been, this was not the central concern of the Iraqi government. Instead, Iraq continued to object to the UNSCOM demand for unlimited access to all sites and returned to the theme that American and British citizens on these teams were engaged in espionage detrimental to Iraq's national security. In particular, Iraq targeted Scott Ritter, the American leader of the team specifically commissioned to uncover the elaborate networks of deception which had allowed Iraq to conceal so much of its weapons programmes from the UN gaze. Despite UN condemnation and barely veiled US threats of punitive military action, Iraq persisted in its stance. This led Butler to spell out in January a plan based on the November UNSCOM panel meeting, whereby advisory committees, which included Iraqi officials, would be established to review evidence of Iraq's development of clandestine weapons. These were intended to determine whether it was possible to close the files on certain categories of weapons. To Baghdad's chagrin, the advisory committee submitted a negative report. In addition, Butler promised greater diversity in the nationalities of the members of these teams and stated that if Iraq were to comply, then sanctions would indeed be lifted.

At the same time the US reinforced its military presence in the region and was backed unequivocally in its position by the UK, which also sent an aircraft carrier to the Gulf. (See map, p. 227.) Inevitably, these apparent preparations for military action galvanised those in the region and on the Security Council who saw the use of force as counterproductive. In particular, Russia renewed its efforts to find a diplomatic solution to the crisis that would satisfy both the UN Security Council and the Iraqi government. Seeing the difficulty faced by the US in rallying international support for any planned military action against Iraq, Saddam Hussein became ever more defiant. He publicly called on the Iraqis to join a *jihad* against the UN sanctions, declared a freeze on all inspections of 'sensitive sites' and renewed his demand that a deadline should be set for the lifting of sanctions, threatening to prevent all further inspections within six months if this did not happen.

By the end of January, Iraq had banned the UNSCOM inspectors from at least eight presidential sites and had also prevented other teams from carrying out their tasks. Meanwhile, the build-up of US and UK forces had continued, backed by warnings from the US President and others that if Iraq continued its defiance of the UN, it would face military action on a scale not witnessed since the Gulf War. The intention was clearly to disabuse the Iraqi leadership of the idea that any military action would be merely symbolic and to suggest that this time it would materially affect the power of Saddam Hussein's regime. Given the nature of his power, it is uncertain that Saddam Hussein was convinced. He could also see that the greater the ostensible military threat from the US, the more opposition to its course of action grew in the region, and the more determined other members of the UN Security Council became to seek a diplomatic solution to the crisis.

Serious efforts were now being made to induce Iraq to accept the principle, once again, of unfettered UNSCOM access. Despite the real danger of military attack which lay behind these efforts, Saddam Hussein calculated that this was all to Iraq's advantage. If an agreement was to emerge, either to pre-empt such an attack or in its aftermath, it would have to gain Iraqi approval, providing Iraq with a degree of legitimate control over UNSCOM that it had not had before and increasing its ability to thwart the inspectors. In addition, although there were some attempts to link such an agreement to more generous oil sales under UNSCR 986, attention had now focused on the need to give Iraq some hope that sanctions would be lifted in the foreseeable future to ensure its cooperation with the weapons inspectors in the present. Iraq had in effect set the terms for its rehabilitation since others were now expending energy and ingenuity in seeking to persuade the country to accept that which had originally been imposed on it.

Throughout February these efforts continued. Russia, France and the Arab League failed to persuade Saddam Hussein to drop his demands that the UNSCOM teams be reformed and that a limited term be set on any inspection of sensitive presidential sites. In the absence of an Iraqi agreement on unlimited and open-ended access, military action by the US and the UK to enforce compliance seemed increasingly likely. It was against this background that the UN Secretary-General, Kofi Annan, flew to Baghdad on 20 February. On 23 February, following a meeting with

Saddam Hussein, Annan and Iraq's Deputy Prime Minister, Tariq Aziz, signed an agreement allowing UNSCOM immediate, unconditional and unrestricted access to all sites. For its part, UNSCOM promised to respect Iraq's national security, sovereignty and dignity. This was to be achieved by forming a special group of diplomats and UNSCOM members, under a commissioner appointed by the Secretary-General, to inspect the eight designated 'presidential sites', whilst in other areas UNSCOM would proceed as it had done before. Most importantly, Saddam Hussein dropped his demand that the inspections should only operate for a fixed term, while the Secretary-General promised to bring the sanctions question before the UN Security Council.

Crisis Averted, But Not Resolved

The Baghdad agreement was approved by the Security Council and for the second time in six months the threat of military action against Iraq had been headed-off. In such circumstances, however, the possibility that it might yet be used remained. All sides drew rather different lessons from the crisis. Iraq predictably portrayed the outcome as a victory for the strategic genius and courage of Saddam Hussein. Others saw it as a salutary example of the power of diplomacy. For their part, the US and the UK, at least, believed that only the threat of military force had convinced Saddam Hussein to back down. Consequently, both the US and the UK maintained their military presence in the region to ensure that Iraq would honour its commitments. Many had reservations about the wisdom of using force in this context. However, others saw the threat of force as vital to resolving such crises and considered that its value when dealing with Saddam Hussein had been confirmed. These mixed signals were contained in a UN Security Council Resolution, backed by the US and the UK and passed on 2 March 1998, which threatened the 'severest consequences' if inspections were inhibited by the Iraqis, but implicitly required such action to be agreed by the Security Council before it could take place.

Saddam Hussein's reasoning in agreeing to end the crisis may have been partly influenced by the threat of force, but he may also have felt that he had already extracted all he needed from the situation for the moment. Not only was his standing greatly enhanced by the special mission of the UN Secretary-General, but his initiatives had brought the link between the weapons inspections and the lifting of sanctions to international attention. It remains in his interest to keep attention focused on that link and to keep up the momentum for the lifting of sanctions, even if that means running the risk of military action. Indeed, he may have seen US acquiescence in the Baghdad agreement as proof of US and British reluctance to use force because of their difficulty in rallying regional or international support for its use.

This hands much of the initiative to Saddam Hussein. Thus emboldened, he may well seek a restrictive reading of the Baghdad agreement, aggravating relations with UNSCOM once again, if it seems that progress towards the lifting of sanctions is too slow or if the teams seem close to uncovering key aspects of Iraq's weapons programmes which are still concealed. In pursuing this strategy, he may calculate that the international repercussions of any US military attack would be to his advantage, outweighing the damage it could inflict on his power in Iraq. This must remain of concern for the immediate future. In the longer term there is now the disturbing prospect that Saddam Hussein will emerge on the other side of the sanctions regime armed with some of the weapons he devoted such massive resources to developing and for the concealment of which Iraq has had to pay such a high price.

New Political Currents in Iran

The unexpected landslide victory of the moderate cleric, Mohammad Khatami, in Iran's presidential election of May 1997 has in many respects transformed the political scene in Iran. The importance of this remarkable victory was further buttressed by two other positive developments. The first was the December 1997 summit meeting of the Organisation of the Islamic Conference (OIC) in Tehran, which for all practical purposes ended Iran's regional and even international isolation. Intensive Iranian efforts to bring all Arab and Islamic states to the conference, particularly those from the Gulf, ensured that it would be a well attended and successful meeting. Even long-term adversaries such as Saudi Arabia and other Arab and Gulf states sent high level representatives, and the conference went a long way to repair Iran's relationship with the Arab world.

The second development was the message contained in Khatami's interview with the American-owned Cable News Network (CNN) in January 1998. This unusual interview was a clear attempt by the Iranian president to reach out to the American people by praising Western culture and American values, while, at the same time, underlining Iran's grievances against the US government. His initial long statement on liberty and freedom as the cornerstone of American society was a theme that had not been heard from an Iranian official since the revolution. Although he did not call for a resumption of official relations, Khatami proposed cultural exchanges between the American and Iranian peoples as a way to bring down the walls of mistrust that had been constructed to separate them.

The full impact of these events on the domestic and foreign policies of the Islamic Republic of Iran will not be apparent for some time. What is clear, however, is that the key episode that made these developments possible was Khatami's presidential victory which turned on two significant underlying factors. First, it affirmed the evolving nature and the growing maturity of Iranian domestic and electoral politics and the masses' yearning for a more open society. Second, and although this was less clearly demonstrated in the electoral contest, it indicated a desire in Iran to move away from the past bellicose, isolationist, and rigid foreign policy positions towards a more inclusive, friendly, and welcoming posture. If sustained, both trends will help alter the Iranian political landscape. Given the country's important strategic position in the Middle East, and its access to energy resources in both the Caspian Sea and the Persian Gulf, a return to moderation on Iran's part will bolster regional stability and help maintain the continuous flow of energy to the outside world for the foreseeable future.

The People Speak Loud and Clear

In the arcane and tightly controlled Iranian electoral system, candidates have to be cleared by a powerful body, the Guardians Council, before they can stand for election. Appointed by the supreme leader, the Council's members have routinely used arbitrary criteria to reject without explanation a large number of potential candidates for elective office. The Council acts as an institutional screening device to eliminate the counter-élite and counter-regime opponents from gaining political power. In the 1997 election, some 238 individuals (including nine women) submitted candidacy applications. Of these only four, none of whom were women, were formally approved.

The regime's professed candidate was the conservative Speaker of the Parliament, Ali Akbar Nateq-Nuri. With his support from the regime, and his high profile, he was far and away the most likely winner. None of the other three candidates were as well-known as the front-runner, but Khatami, whom the ruling clique had clearly only allowed to run for election because they expected that he would be easily defeated, had some advantages. He was known to an important segment of the population, mostly because of his service for some ten years (1982–92) at the helm of the powerful Ministry of Culture and Islamic Guidance. He was admired for his role in relaxing censorship and allowing for less restrained print, film, and media expression than other holders of the same position. Khatami's ministerial tenure had made him especially popular with the intelligentsia, women, and younger people. That his more liberal ideas had eventually run foul of the conservatives in the Parliament and he was pressured to resign was a fact well known to his supporters.

Khatami was campaigning on a popular platform. His messages included distinct references to, and celebration of, the rule of law; support for the interests of civil society; and promotion of pluralism, national pride, and non-rigid and inclusive Islamic dogma. He also propounded respect for Western values such as individual freedom, dialogue with the West, economic self-reliance, and social justice. A cynic could view these as no more than campaign promises. What was important, however, was that Khatami's stance on openness and the rule of law was believed and received enthusiastically by the people.

The apparently free electoral process allowed for a significant and forceful expression of popular sentiments. This was one of those rare occasions in the Middle East where voters actually believed in the efficacy of their ballot box decision. Those eligible to vote noted that, unusually, there were indeed serious differences and a real choice between the two leading candidates, and they participated on a massive scale.

As Khatami's agenda became better known – and as he engaged in an intensive Western-style campaign, including opening campaign head-quarters in all corners of the country, frequent travel and speaking engagements, three nationally-broadcast debates and a web site – his popularity and appeal increased dramatically. Opinion polls pointed to a rise in Khatami's popularity from some 14% about a week before the election to 52% a few days later. In the election itself some 29 million voters (89% of the eligible electorate) took part, and over 20m (close to 70%) voted for Khatami. The people had spoken thunderously, and the message was one that the ruling clique could neither have expected nor wanted to hear.

It Was No Fluke

Khatami had, and retains, strong support from across the social spectrum but particularly form younger people and women. The youth vote was overwhelmingly for Khatami. Two-thirds of the country's population is under the age of 25 and, since the voting age in Iran is 15, a substantial number of voters were born after the Islamic revolution of 1979. The young people's vote for Khatami was to some extent a symbolic protest against the regime's revolutionary rhetoric and rigidities, its cultural and educational restrictions, and the direction of its foreign policy. This was a vote for a breath of fresh air, a more open culture, and less interference by the state in people's private affairs.

Women also voted in huge numbers for Khatami because they saw him as a man who would support and extend the important reforms affecting the position of women that had been slowly gaining momentum. The treatment of women in revolutionary Iran has gone through three distinct stages, from the oppression of the early days of the revolution, through a somewhat more moderate phase a year or two into the Iran–Iraq war, to a far more liberal approach since the end of the war and the death of Ayatollah Khomeini in 1989.

Much of the credit for the improvement is due to the struggles of women activists within Iran. Although women are still repressed in Iran, at least in comparison with women in the West, their position has improved in important ways since the revolution and compares favourably with that of women in the Persian Gulf states and much of the Middle East. The election gave women an opportunity to express the strength of their concern that the gains in legal, educational, cultural, and other critical areas should be preserved and expanded. Presented at last with a candidate who stood out clearly as a likely champion, they voted for Khatami in droves.

Otherwise, support for Khatami came from a wide and diverse range of interest groups. It appears that a large number of middle class voters, civil servants, and the intelligentsia voted for him, although it is hard to assess the full extent of his support from this sector. Support from the more secular-minded intelligentsia was to be expected since Khatami promoted a degree of cultural openness and had developed a reputation for tolerance while he was the Minister of Culture and Islamic Guidance. The vote for Khatami from some members of the Islamic intelligentsia, an integral part of the Islamic system, adds an important dimension to his wide appeal. That vote probably reflected not so much enthusiasm for Khatami's policies, however, as the divisions in their ranks and tensions over the issue of the proper place of clerics in public and political life, particularly the role of the all powerful institution of *velayat-e faqih* (rule of supreme jurisconsult). The controversies that have engulfed the Iranian Islamic modernist, Abdolkarim Soroush and similar activists have also caused rifts.

Finally, Khatami also received support from other sections of society including the poor and the underclass. Again, it is hard to be sure just how much of his vote came from this direction since no exit polls were conducted. It is clear, however, that given the uneven performance of the economy and the regime's failure to achieve its lofty goals of establishing social justice, the poor and the underclass expressed their discontent through their votes.

Forces of Opposition

As important as Khatami's extensive, yet amorphous, constituency was, it cannot sustain itself indefinitely. The voters were mobilised effectively for the election but this is a far cry from being organised in institutional forms such as political parties. Moreover, there are important and established forces of opposition that will constrain Khatami's programmes and actions, especially in the foreign policy arena.

The forces of opposition, or potential opposition, are made up of a disparate but powerful set of institutions and actors with entrenched political, economic, and ideological interests. While aware of Khatami's massive electoral victory and popular support, they can find ways to thwart his reform agenda. The most important of these opposition forces can be found in institutions such the current parliament, the state-supported foundations, segments of the bazaar, and elements within the military/ intelligence complex that encompasses the Revolutionary Guards and the Interior and Intelligence Ministries. Moreover, Khatami's decisions can be constrained (if not overturned), particularly in the foreign policy arena, by the preferences of the supreme leader, Ayatollah Ali Khamenei.

The current parliament has a conservative majority and is headed by Khatami's defeated rival, Nateq-Nuri. It is made up of a wide array of factions, which are loosely grouped along interest and ideological lines. In spite of the conservative majority, Khatami was able to obtain parliamentary approval for every one of his ministers, including at least four controversial liberal nominees. This was his first major victory after the election and demonstrates his political acumen and ability to get results within the system. Moreover, the next parliamentary elections may produce a legislative body far more supportive of Khatami's programme than the current group.

The foundations (*bonyads*) are a significant economic and political force in revolutionary Iran. Although philanthropic foundations with religious backing are nothing new in the Muslim world, a special set of foundations that were created in Iran after the revolution are unusual. These were set up by the Islamic government with funds confiscated from the Pahlavi family and close affiliates of the old regime and other forms of state assistance.

Two of these foundations – Foundation for the Oppressed and Foundation for the Martyrs – were created to assist the underclass and the families of those who died or were injured in the revolution and the Iran–Iraq War. The supreme leader appoints the heads of these organisations, whose enormous resources of many billions of dollars have made them critical players in the Iranian scene. Their roles have been buttressed by their immense powers of patronage and the strong patron–client relationships they have established. Given that the sympathies of the foundations' leadership do not lie with Khatami – they supported his opponent in the election – they may well emerge as significant institutional opponents of the president. Although their opposition will not necessarily be expressed openly, they can exert significant pressures behind the scenes. They can particularly be expected to act if Khatami tries to put in practice economic reforms which would restrict some of the often monopolistic practices of these institutions.

The bazaar is another potential problem area for the new regime. This complex, difficult-to-define community, has always played an important, and at times vital, role in Iranian politics. The bazaar merchants are extremely protective of their economic prerogatives and their pivotal role in the Iranian economy, especially in foreign trade. Khatami does enjoy some support among segments of the bazaar but he will be opposed by the largest part of the bazaar community if he attempts to bring large foreign investments into Iran, thus jeopardising their interests.

The military and intelligence complex is another serious potential opposition. Its preference for Khatami's opponent became apparent very early in the election. This entrenched group enjoys immense power in Iran and is tied very strongly to the supreme leader. Khatami was aware of the dangers and one of his early achievements as president was his replacement of the head of the Revolutionary Guards and the Minister of Intelligence. The latter's reputation had been badly tainted through his role in the acts of terrorism in Germany against opponents of the Iranian regime. Moreover, Khatami named, and received parliamentary approval for, a moderate Minister of Interior, whose duties include supervising the national police and the electoral process. Yet, even though the new president has been able to place some of his own people in key ministries and organisations, the actual power of the newcomers is far more limited than hoped. The entrenched upper echelons of the bureaucracy in these institutions, who are often ideologically driven, will not easily relinquish their privileged position and accumulated interests.

Last, but by no means least, is the key position of the *Wali Faqih* (supreme leader), Ayatollah Khamenei. The person who holds this extremely powerful constitutional position has near dictatorial powers, with the final word on intelligence and security issues and on Iran's foreign policy. Khamenei has been careful not to appear overly antagonistic to the popular president. He has, nevertheless, made speeches and taken positions that are not in conformity with those of Khatami. The leader's unexpected opening statement at the Organisation of the Islamic Conference was rigid and harsh, in sharp contrast to Khatami's positive and tolerant succeeding presentation. Likewise, Khamenei's speech, soon after Khatami's mild message to the American people, was vindictive in tone and strongly critical of the US. Both of Khamenei's statements appeared as atavistic throwbacks to the earlier revolutionary zeal of the Khomeini era.

Future Directions

The process of internal reform, particularly when pitted against an entrenched ideological regime, can easily be reversed. There is no particular reason to suppose that the situation in Iran is any exception. Yet the changes that have occurred in the domestic scene, and especially Khatami's victory, have given citizens a sense of empowerment which they will not easily relinquish. Although a tug-of-war between the reformers and the hardliners can be expected, the prospects for further reform in Iranian domestic politics are reasonably good. The fundamental test will be the extent to which Khatami is able to gain control over vigilante activities by the Islamic ideologues, *Ansar-e Hizbollah,* and other hardliners in the system. Control over these groups would go a long way to ensure respect for the rule of law and protection of the interests of a civil society.

The situation is perhaps even more complex in the foreign policy arena. The Iranian government has made a serious and concerted effort to patch up its differences with the Arab world. This has been largely successful, particularly with regard to Saudi Arabia, Jordan, Qatar, Oman and Kuwait. The rapprochement with Saudi Arabia accelerated in June 1997, bolstered by a common concern over Israel's settlement building in Jerusalem and the Occupied Territories. It continued through the participation of Prince Abdullah Bin Abdul-Aziz, the crown prince and deputy premier, and Prince Saud al-Faisal, Saudi Arabia's foreign minister in the 54 country Islamic Conference hosted by Iran in December 1977. In February 1998, former Iranian President Rafsanjani led a high level delegation to Riyadh where he told Saudi Arabia's consultative council that 'we look to Saudi Arabia as the *qibla* (prayer direction) for all Moslems, as the cradle of Islam', thus accepting a Saudi position that it should retain sole guardianship of the Islamic holy shrines of Mecca and Medina.

The effort has been far less successful with Egypt as no formal diplomatic relations between the two countries even exist. Iranian leaders still find it difficult to accept the peace between Egypt and Israel, and there are problems that are difficult to overcome with regard to the leading role which Egypt wishes to play in the region. Yet none of this prevented the Egyptian foreign minister from attending the OIC meeting in Tehran. Iran continues to have problems with the United Arab Emirates, in spite of extensive economic and trade relations, as the dispute over the three Gulf islands (Greater and Lesser Tunb and Abu Musa), which both claim, lingers on with no resolution in sight. On the whole, however, Iran's relations with the Arab and Islamic worlds have improved considerably and can now be considered quite positive.

Relations with Israel remain at loggerheads and can be expected to stay that way for the foreseeable future. Even so, there have been many rumours and reports of secret contacts, agricultural exchanges, and other clandestine relations. The prospect of better relations between the two countries would improve if the thorny issue of Iranian support for the Lebanese *Hizbollah* could be resolved. But Iran will undoubtedly continue unabated support for them until Israel withdraws from southern Lebanon and this aspect of the Arab–Israeli conflict is resolved.

Relations with the West are also complicated. Although serious disputes with European Union countries have been partly settled, the interrupted critical dialogue has not resumed. Moreover, the Salman

Rushdie case continues to be considered a serious human rights issue in Europe and a major obstacle to dialogue. On the more positive side, significant economic contracts with major European (and Russian) energy companies have been signed in spite of the American embargo. With Khatami's election victory, there is an opportunity for a gradual expansion of the positive trends.

The litmus test of Iran's improved relations with the West will centre on the US. Iran is a long-term challenge for US foreign policy. Since the Iranian revolution, US policy towards Iran has, at best, had only limited success. The latest policy, that of 'dual containment', with extra-territorial sanctions and an embargo on Iran, has not fared much better. Serious disputes between the two sides exist over US charges of Iranian terrorism, violation of human rights, development of weapons of mass destruction, interference in the internal politics of Arab states, opposition to the Arab–Israeli peace process, and an armament build-up. The tone of these charges and counter-charges has eased somewhat since Khatami's election. Although there has been no breakthrough, the US has made it clear that it is hopeful that the Khatami government can follow its more moderate talk with more moderate policies. No one was talking about government to government talks, but expansion of the minimal cultural exchanges can be expected. In February 1998, a US wrestling team was warmly welcomed in Tehran when it took part in an international competition. Unless serious domestic political problems emerge in Iran, the prospects for a reduction of tensions between the United States and Iran are more promising than they have been at any time since the Shah was overthrown.

◆

Asia

Many Asians will prefer to forget 1997, but it was a year which will have long-term consequences. Most of the news was bad, but there was a small piece of good news to go with it. The bad news was that the deepest economic crisis since East Asia's economic boom began decades ago caused a financial collapse that will take billions of dollars from outside the region to repair. The small item of good news was that Asian governments began to confront the fact that change is necessary if economic growth is to be restored. With all this upheaval, security issues were of secondary importance, but if economic and political reforms are not undertaken, there could be serious military implications.

The Asian economic contagion began with the collapse of the Thai currency on 2 July 1997. The infection devastated Indonesia and laid low even the relatively mature tiger economy of South Korea. No country in Asia escaped entirely, although China, Taiwan and Japan have not yet seen a significant decline in growth. Others in Asia, most notably North Korea, were already in dire economic trouble, and even the once surging India began to falter. No wonder there was economic gloom and the cheery talk of 'strong economic fundamentals' had faded. There was no sign of brighter prospects in the immediate future.

The political consequences remain unclear. South Korea and Thailand responded to the economic crisis with a change of government and a new determination to undertake long-term and far-reaching reforms. Their struggles would be protracted, but the first step – recognising the link between political and economic reform – had been taken. India also had a new government, but not because of the crisis. Here, the result looked like being a setback for economic reform. Most attention focused on Indonesia – the linchpin of South-east Asia, but a country with a determined 'one-family' government resistant to reform.

With Indonesia in denial, so that its recovery was highly uncertain, no one can be optimistic about a speedy improvement in South-east Asian fortunes. Important questions also envelop Japan and China. With the political and economic system stalled in Tokyo, and Beijing struggling to keep its economy out of recession while undertaking sweeping economic reform, there could be no confidence that the powers of the region would lead the recovery. The crisis showed Asia not only to be without its much-vaunted new form of capitalism, but also to be without indigenous leadership. The most optimistic view of the region can only be that the worst is probably over.

China: Slowly Joining The World

During 1997, China's commitment to becoming a more integral part of the international community became clearer than before. A major indicator was the smooth transfer of sovereignty over Hong Kong on 1 July, followed by China's close adherence to its long-standing commitment to allow its people to exercise a high degree of autonomy. An indication of the deepening engagement was provided by President Jiang Zemin's state visit to the United States in October–November. But perhaps the most significant sign of how far the process has developed was the response of China's leaders to the Asian economic crisis.

It would have been easy for Chinese leaders to see the Asian crisis as a warning against further reform and opening of the economy. China, after all, was sheltered from the immediate effects by the non-convertibility of its currency and by the various barriers that still screen China from full exposure to the international economy. Instead, China's new generation of leaders recognised that it was a timely warning for them too. Hence, at both the Party Congress in September and at the National People's Congress (NPC) in March, the leadership tied their colours firmly to the mast of structural reform. It recognised that a fundamental makeover was needed for the old Soviet style state-owned enterprises (SOEs) and that the financial sector had to be put on a modern footing and released from the growing and unsustainable burden of the bad debts incurred by the SOEs.

China's leaders earned considerable good will by agreeing, soon after the crisis broke, to provide $1 billion as part of the rescue package for Thailand. They were also widely praised for their frequently stated public commitment not to devalue their currency, as that would have compounded the economic instability in the region and possibly set off a series of competitive devaluations. To be sure, the decision served Chinese interests too, but it illustrated both the extent of China's interdependence with the regional economies and its acceptance of a new kind of responsibility which grew out of that interdependence.

The deepening engagement did not suddenly dissipate the longer term uncertainties and suspicions engendered by China's steady rise. But it did suggest that a more firmly engaged China might be easier to deal with. It did much to confirm the view that the process of reform in the country was irreversible, despite the social costs, for example in unemployment. China's leaders appeared to be more attentive to the concerns of others and seemed increasingly to be adopting the conventions of international diplomatic practice. Thus, over the last two years, Beijing has signed a number of international agreements and conventions, such as the Comprehensive Test Ban Treaty (CTBT) and the International Convention on Social and Economic Rights, that entail a degree of cost. It ratified the Chemical

Weapons Convention, thus accepting for the first time international inspections on its territory. Since the hand over of Hong Kong, it has undertaken, on its own initiative, to report to the UN on Hong Kong's adherence to the two human rights conventions, despite having fiercely resisted British attempts to persuade the Chinese side to do so in the course of their negotiations.

China's leaders have also shown an ability to use the new style of multilateral diplomacy to pursue goals that implicitly challenge the existing basis of order in the region. In March 1997, China chose the multilateral setting of the Association of South-east Asian Nations (ASEAN) Regional Forum to argue that the bilateral military alliances in Asia were outdated relics of the Cold War and should be superseded by the new modes of cooperative security. Americans who were present were discomfited by the unwillingness of their Asian allies to challenge the Chinese view openly, even though they and their Asian colleagues privately insisted that their security was closely tied to a continued US military presence in the region. This episode suggested that problems posed by the rise of China involve more than the question of how far the Chinese adhere to accepted international norms and practices. Long standing issues remain, including, *inter alia,* the Taiwan problem, the competing territorial and maritime claims in East Asia and, more broadly, the kind of balance that can be drawn between Tokyo, Washington and Beijing. All extend the China problem into areas beyond the issue of engagement itself.

A New Technocratic Leadership

A China that is more closely engaged with the international community is one that should be easier for that community to deal with in the longer term. While its transition is under way from a highly personalised system of rule to one in which institutions will play a greater role, even at the highest level, much will still depend upon the character of the top leadership. The succession of Jiang Zemin as the core leader had already taken place before the enfeebled Deng Xiaoping eventually died in February 1997. Many inside and outside China were surprised by the ease and speed with which he settled in after the demise of the once paramount leader. Jiang had been seen as a transitory figure when he was first picked out by Deng shortly after the Tiananmen Square killings in June 1989. But without appearing to be particularly accomplished, he has steadily strengthened his position over the years through careful appointments and skilful manipulation of divisions within the Party.

Most of the key positions in the military in the late 1990s are held by generals Jiang has appointed or promoted, and he has been able to bring into the political bureau key figures from his own so-called Shanghai faction. In the spring of 1997 he consolidated his position by seeing-off the remaining leftists in the party so that the country could proceed with the much delayed restructuring of the SOEs. By the autumn he was able to ease out his main competitor, Qiao Shi, who, favouring a degree of political reform, occupied a position more to the right of the political spectrum. Jiang was able to oust him from the Standing Committee of the Political Bureau at the 15th Party Congress that was held in September. Despite these successes, Jiang has not been able to exercise full dominance and he has to rule by consensus and by balancing divergent interests.

Clearly, Jiang Zemin does not exercise the same kind of personal authority as his predecessor, or indeed of the other elderly founding fathers. Nor would any other member of the present leadership be able to. The old guard were stalwarts of the revolutionary wars and struggles that predated the establishment of the communist state, and had commanded armies as well as presiding over bureaucracies. They drew their power from their personal standing rather than from their institutional positions. Jiang was described by Deng as the 'core of the third echelon' of leaders. This is the generation that joined the Party in the 1940s and received Soviet-type training in the 1950s. Jiang and his associates were mainly educated as engineers; they gained experience in the state-owned sector, first as managers and then as bureaucrats. Although better educated than their seniors, they seem to lack their breadth of vision and ideological conviction. They may be best understood as a Soviet-type technocratic élite, albeit with Chinese characteristics.

What has distinguished Jiang is not so much that as the head of the Military Affairs Commission (the supreme body in charge of the military) he is a civilian with no military experience, but that he is the only civilian on the Commission. The corollary is that the 15th Party Congress in September, unlike its predecessor, did not nominate anyone from the military to sit on the Party's Standing Committee of the Political Bureau. This signalled a clear intention to designate the People's Liberation Army (PLA) as an exclusively professional military body. It did not mean that the military had suddenly ceased to be of major political significance, but it indicated an agreement between Jiang and his civil colleagues, on the one side, and the senior military personnel, on the other, that the PLA should be less involved in day-to-day politics and that it should focus on its own modernisation.

A More Accommodating China

Following the adverse reactions to its assertiveness in the mid-1990s, when it established a presence on Mischief Reef in the South China Sea and tried to intimidate Taiwan with missile tests in the seas off Taiwan's two main ports, China has recently been pursuing more accommodating policies. It signed the CTBT in 1996, even though elements within its military establishment argued that this would place China at a disadvantage, since the technologically more advanced nuclear powers would be able to continue to carry out a wider range of experimentation. The multilateral treaty that China signed in 1996 with Russia and the adjacent three Central Asian Republics not only helped to stabilise the border areas through a variety of confidence-building measures, but also confirmed Chinese acceptance of the multilateral format, as opposed to their previous exclusive preference for unilateral action and bilateral diplomacy. They appeared to become more comfortable in dealing with the multilateral format of the ASEAN Regional Forum. In fact, by March 1997 the Chinese showed that they could use multilateralism as a new means by which to signal their opposition to what they perceived as US hegemony. Yet they did not press the issue, which arose particularly from their fears that the new Japan–US security guidelines would have authorised Japan, under certain circumstances, to extend its area of military operations to the Taiwan Straits. Instead, the Chinese appeared to have been mollified in the course of a flurry of high level official exchanges, including between the military.

The return of Hong Kong to Chinese sovereignty on 1 July, which was universally regarded as a major test of China's adaptability and of its professed good international citizenship, passed without major incident. If many in Hong Kong and the world beyond feared Chinese interference in the new Special Administrative Region (HKSAR), China's leaders feared trouble would be instigated by the departing British or be caused by the unnamed 'dark foreign forces' that the more conspiratorially minded Chinese communists profess to see as always planning subversion. In the event none of these fears materialised.

Far from interfering, Beijing, having ensured the appointment of a few key leaders, allowed the HKSAR to conduct its affairs very much as it had before the change in sovereignty. Even in matters of foreign affairs, where the Chinese had exclusive powers according to the Sino-British Joint Declaration, the Chinese Foreign Commissioner in Hong Kong announced on two separate occasions that, on Beijing's instructions, several matters that could have come under his umbrella would instead be placed under the jurisdiction of the HKSAR as part of its exercise of foreign economic relations. The judiciary and the civil service acted very much as before, except that the Provisional Legislative Council was considerably less effective than the legislature that had been elected under the Patten rules. The electoral system selected for the first Legislative Council of the HKSAR has been generally criticised as unfair and designed to ensure that the executive will have a built-in majority. Nevertheless, the legislature to be elected in May 1998 will at least have the merit of having some representatives of the Hong Kong democrats. Even the media, which continued to be somewhat circumspect in its handling of China-related news, became ever more censorious of the local administration, especially with regard to the handling of the 'bird flu'.

Instead, Hong Kong was assailed by the one crisis that was not anticipated and that could not be blamed on either the British or the Chinese – the Asian economic crisis. The crisis threatened to spread to Hong Kong in October when speculation developed against the peg tying the HK dollar to the US dollar. The defence of the peg was successful, but at the cost of high interest rates and a fall in the property market and the stock exchange. The Beijing government stood by the HKSAR government and intimated that, if necessary, it would use its vast foreign currency reserves, then at around $130bn, to back up those of HK which were about $80bn.

According to the out-going Premier Li Peng, the highpoint of 1997, which was deemed a good year for China overall, was Jiang Zemin's state visit to the US in late October and early November. The Chinese saw this as indicating firstly, that the Americans were no longer seeking to contain China, and, secondly, that they were committed to establishing a 'strategic partnership' with China into the next century, despite the differences over Taiwan, human rights and Tibet which remained. Interestingly, the Chinese began to tone down their earlier criticisms of the new strategic guidelines agreed between the US and Japan, which they saw as designed to enable Japan to play a more active role in the Taiwan Straits. The better understanding with the US also seemed to clear the way towards a resumption of the negotiations between Taipei and Beijing that were suspended after Lee Teng-hui's visit to the US in June 1995. By March 1998 each side was edging in that direction despite their deep mutual distrust.

Jiang's visit coincided with a flurry of high level meetings between the leaders of China and Russia; Japan and China; Russia and Japan and Russia and the US. Official Chinese commentaries depicted these as demonstrating a new pattern of relations between the great powers. These were now seen as defined by common interests rather than alliances; by dialogue replacing antagonism in solving disputes; and by establishing relations that are explicitly not aimed at third countries. All in all, the new relations were seen as establishing a trend towards multipolarity among the great powers which also included France, Germany and the UK. But perhaps the key issue from a Chinese perspective was what the commentators saw as the common reluctance of these large powers to be 'subjected to only one voice', namely that of the US.

Grasping The Nettle Of Economic Reform

Both the 15th Party Congress in September 1997 and the 9th National People's Congress in March 1998 put the emphasis on significant economic reform, even at the cost of what was termed short-term pain. A clear decision had been taken: the economy could no longer continue to support the huge loss-making SOEs. Chinese statistics are notoriously unreliable, but there can be little doubt that the economy has been slowing down since the double digit growth rates achieved between 1992 and 1995. According to official figures, the annual rate of growth fell to 9.6% in 1996 and 8.8% in 1997. It is expected to reach 8% in 1998. Even the Chinese concede that the figure is suspect in that, since it is a measure of output, it includes the inventories of unsold production that have been accumulating in many SOEs. Some external analysts calculate that the true growth rate could be as low as 4%, although most estimate that it will be 6–7%. At the same time, China's state banks have been pouring money into SOEs and into ill-conceived construction projects. According to the People's Bank of China, the bad loans amount to 5–6% of total loans – three times more than previously admitted. According to *The Economist* in February 1998, at least 20% of the total loans are non-performing by international standards, the equivalent of $145bn at the end of 1996 (or 18% of gross domestic product). Since the declared capital base is just $54bn, by any criteria other than their own, China's state banks would appear to be bankrupt.

The initial plan of the September Party Congress was to support between 500 and 1,000 large SOEs and encourage the remaining small and medium sized enterprises to find their own way to solvency through mergers, establishing companies, joint ventures and what would be described in the West as forms of privatisation. The very worst cases would be allowed to go bankrupt. The SOEs employ between 75 and 100 million workers, the bulk of the urban work force, and the Minister of Labour has estimated that 10m of them would become unemployed by the end of 1998. The potential for social disruption is immense as, in true Soviet fashion, the SOEs were not just units of production, but provided housing and all forms of social security to their workers who thought they enjoyed lifetime employment. Only a few select localities have an effective social security net in place to help those made redundant. Meanwhile, the labour market increases naturally year by year as 13–18m people come of age across the country. If the so-called floating population, estimated at 70–130m, who have left the countryside in search of urban work, are included in the equation, it is clear that maintaining a high rate of growth in the Chinese economy is vital.

As a factor that could slow economic growth, the Asian economic crisis has considerable significance for China. Much of the past high growth in the Chinese economy stemmed from the rapid increase of exports and massive foreign investment. Both are expected to drop significantly in the second half of 1998 and into 1999. This is happening at a time when domestic demand has slowed to the point where inflation rates were negative towards the end of 1997.

The leaders expect to be able to reinvigorate the Chinese economy through huge infrastructure investments. In February 1998 it was announced that up to $750bn would be invested in power plants, roads, rail, bridges and other infrastructure over the next three years. The following month, at the NPC, the figure was raised to $1 trillion. How these vast sums will be raised has not been stated, but such is the high rate of family savings that the government could reach its goals if it were to find the means to direct these savings where it chose. That, however, may be easier said than done. Meanwhile, as Zhu Rongji assumed the premiership in March 1998, ambitious plans were announced to increase administrative efficiency by streamlining government ministries and commissions and sacking up to 4m of the 8m officials employed by the central government, with further sackings to follow as the efficiency drive reaches the provincial level.

The military have not fared too badly in the general shake-up, despite being required at the Party Congress to reduce their ranks by a further half million from the current level of around 3m. Their nominal budget is to be increased by 12% to reach $11bn making this the largest single item of budgetary expenditure, with education, for example, being allocated only the equivalent of $1.2bn. Moreover, as part of an ambitious plan to raise the country to the rank of second in the world in terms of advanced technology over the next 20 years, the military-related industries have been singled out for special attention. One of the new so-called super ministries is to be called the State Commission of Science, Technology and Industry for National Defence.

As China deepens its engagement with the outside world, it is apparent that its key problems remain domestic and much will depend upon its capacity to carry out the structural reforms which Zhu Rhongji has championed and which can no longer be delayed. That in turn will depend on maintaining sufficiently high economic growth to ensure that the ranks of the unemployed do not become so numerous as to threaten social and political instability. To avoid any new foreign problems arising while it concentrates on solving its domestic difficulties, China's accommodation to the Western world, and the US in particular, can be expected to continue, but in a guarded fashion.

---◆

Dark Times In Japan

A year that started off well for Prime Minister Ryutaro Hashimoto – seemingly secure in his position and increasingly accepted on the international circuit as a determined leader – turned sour as the year wore on. Japan proved no longer immune to the contagion of economic and financial crises that spread across the Asia-Pacific region. Its own financial problems burst into the open with a succession of bank failures and finally the collapse of the country's fourth-largest brokerage house, the largest corporate bankruptcy in post-war Japanese history. Japan's extensive trade, aid and investment links with the Asia-Pacific region meant that it could not avoid being affected by events elsewhere in the region.

Hashimoto indeed seemed to be losing his touch by late 1997. His uncertainty over how to deal with the financial crises and political miscalculation in his cabinet reshuffle outweighed his attempts to promote a more active image in international summitry and to introduce his long-awaited administrative reform programme. By early 1998 his popularity had sunk to its lowest level since he took over as prime minister. Even though the disintegration of the major opposition party seemed to offer him some respite, and potentially set up his Liberal Democratic Party (LDP) to regain its majority in forthcoming Upper House elections in the summer of 1998, Hashimoto remained a troubled man.

Japan, too, remained troubled. The world's second largest economy suffered its darkest year this decade. Vain efforts by companies and the government to play down the severity of the economic problems (which stemmed from the bursting of the speculative 'bubble economy' at the beginning of the 1990s) finally came home. Nevertheless, although many Japanese felt resentful about losing their investments and even their jobs, no major cracks have appeared yet in Japan's much vaunted social stability.

Edging Back To A Majority

The elections in October 1996 seemed to bring order back to the Japanese political scene after three years of unstable and short-lived coalition governments. The LDP, led by Prime Minister Hashimoto, regained its old position of sole power even though it did not have an outright majority. The leading opposition party, the New Frontier Party (NFP), led by Ichiro Ozawa, one of the most ambitious of the younger generation of leaders, failed to make as much ground as had been expected.

Worse followed for Ozawa in January 1997, as one of his political colleagues, former prime minister Tsutomu Hata, took a dozen followers out of the NFP to set up another new party, the *Taiyo* (or Sun) Party. Another former prime minister, Morihiro Hosokawa, left soon after. Criticism of Ozawa's leadership style increased within the NFP and pressure for him to step down began to build. But in the spring of 1997, Ozawa was handed a temporary life-line, ironically by his rival Hashimoto.

Despite being by far the largest party, the LDP was just short of an overall majority in the Lower House and well short of a majority in the Upper House. To pass legislation it has had to rely since October 1996 on working agreements with its former coalition partners, the Social Democratic Party of Japan (SDPJ), which had been reduced to a small force in the 1996 elections, and the minuscule *Sakigake* party. But in the spring of 1997, the LDP had to find new partners for the SDPJ, traditionally opposed to the security alliance with the US, balked at supporting the LDP's bill to override local objections and renew the leases for the US bases in Okinawa. One option was the third force in Japanese politics, the Democratic Party (DP), but its co-leaders were split over how to respond – one in favour and the other opposed to joining with the LDP.

Hashimoto turned to Ozawa, who has always supported a strong role in international affairs for Japan. Two lengthy evenings of negotiating achieved the support of Ozawa and the NFP and, at the same time, revived Ozawa's position as someone of influence in domestic politics. On 11 April 1997 the bill passed through the Lower House with the NFP and the majority of the DP supporting the LDP. It passed the Upper House before Hashimoto set off on a trip to Washington.

Hashimoto's success in harnessing support from the NFP and the DP has led some LDP strategists, including the still influential former prime minister Yasuhiro Nakasone, to begin talking about a new conservative grand alliance. The NFP was put together in 1994 from four middle-sized parties and a number of smaller groups, but around one-third of its politicians, including Ozawa himself, were originally members of the LDP. Hashimoto and his close advisers calculate that many NFP members might be lured back into the LDP fold on a more permanent basis.

However, another group within the LDP, led by the party's Secretary-General Koichi Kato, do not trust Ozawa and the ex-LDP members of the NFP. They argue that the LDP should stick with the SDPJ and *Sakigake* and try harder to woo the more liberal wing of the DP. The resultant impasse between these different advocates within the LDP meant that, for all Nakasone's talk, any new plans for political realignment stalled during the summer and autumn of 1997.

Nevertheless Hashimoto's position improved steadily as NFP members, one by one, left to join the LDP. By early September, for the first time in four years, the LDP secured an outright majority in the Lower House. The same month, Hashimoto was re-elected unopposed as the leader of the LDP, thus securing his position as prime minister. He became the first LDP president since Nakasone, 13 years earlier, to receive a second two-year term in office. With Hashimoto apparently making progress both on his cherished plans for administrative and financial reform and on the review of the guidelines for Japan–US security cooperation, there were no rival candidates within the LDP in a position to make a move against him.

Hashimoto's triumph was short-lived. As part of the almost obligatory cabinet reshuffle following his re-election, he brought Koko Sato into his cabinet. Sato is an elderly politician, convicted of having accepted bribes as part of the Lockheed scandal in the early 1970s, and his entry into government provoked widespread criticism. Although Hashimoto argued that Sato was the kind of tough politician he needed to push through his administrative reform programme, Sato's inclusion had more to do with Hashimoto's desire to balance the factional groups within the LDP. Hashimoto's new-found over-confidence probably also played a part. Within days, he was forced to back-track and replace Sato. This debacle was probably still in voters' minds in October, when local elections brought further setbacks for the LDP. In the conservative stronghold of Miyagi prefecture, the LDP-backed candidate lost the gubernatorial election, while in two large cities LDP-backed mayors barely scraped in over communist-backed candidates.

Goodbye To Complacency ?

Political blunders were not Hashimoto's only problem. By late 1997 the economic and financial crisis facing Japan, partly of its own making, partly the result of the financial turmoil spreading across South-east Asia, was also beginning to make serious inroads into the credibility of the Hashimoto administration.

After some signs of recovery in 1996, when economic growth actually exceeded 3% for the first time for five years, 1997 proved to be a year of painful setbacks. The rise in the consumption tax in April depressed consumer spending more than anticipated and effectively strangled a nascent recovery. Instead, in the second quarter of 1997 the economy suffered its most serious contraction for 23 years; over the year as a whole, gross domestic product (GDP) grew by less than 1%. Investor confidence in the Tokyo Stock Exchange slumped following revelations that leading securities houses had engineered windfall profits for favoured clients, among them members of crime syndicates. Their losses were hidden away by switching them between various companies with different accounting periods – with the collusion of Ministry of Finance (MOF) officials. In 1997, these losses finally caught up with the oldest of the 'big four' brokerages, Yamaichi Securities, which went out of business in November leaving behind debts of ¥3.2 trillion ($25 billion).

A number of banks and insurance companies could no longer cope with the huge bad debts they had accumulated on their balance sheets and a succession of regional banks led by Hokkaido Takushoku, one of Japan's ten largest banks, went bankrupt during the year. Other sectors were hit too, especially construction and retail companies; corporate bankruptcies reached a new record high. Even ever-optimistic government officials admitted that the economy was at a standstill by the autumn.

Disregarding the obvious need to reinvigorate the economy, the government continued to try to repair the huge budget deficit, which had ballooned in previous years as the result of successive government-backed stimulation packages. In November, Hashimoto introduced legislation that would curtail the government deficit to less than 3% of gross domestic product by the year 2004 (in 1997, it was 6%, the second highest level in the industrialised world after Italy). The budget-saving measures threatened to wound the economy further. Hashimoto himself then added to the turmoil by reversing policy three times in as many days as he agonised over whether to use public funds to bail out Japan's shaky banking system, before ultimately deciding in favour. Finally, in mid-December, Hashimoto introduced a surprise ¥2 trillion ($15.5bn) stimulus package in which income and residential taxes were cut in order to boost domestic demand, although this was directly at odds with his declared aim of balancing the budget.

Hashimoto had been encouraged to take this step during his discussions with the leaders of South-east Asia, China and South Korea at a summit meeting of the Association of South-east Asian Nations (ASEAN) in Kuala Lumpur. This meeting had helped bring home to him the severity of the economic crisis in the region and how closely Japan's fortunes were tied up with regional problems. Japanese banks have been lending heavily to Asia in recent years; they have an estimated $24bn exposure in South Korea and $37bn in Thailand. Slower growth throughout the region has affected Japan's exports significantly, while plummeting exchange rates mean exports from the region threaten to undercut Japanese exports in major external markets such as the US and Europe.

When Hashimoto sought to convince his political allies that the December tax cuts were absolutely necessary, he argued that he was putting his political life on the line. The government may not succeed, but, by early 1998, it had abandoned its early complacency and had begun to tackle the principal issues.

New Security Guidelines

Hashimoto undoubtedly grew in confidence in his international dealings during the year, but the results were mixed and did not compensate for the domestic problems he faced. The restoration of security issues to the centre of the relationship with the US was a major feature of 1997. While Japan–US trade disagreements continued to be contentious, and even reached the stage of US threats to impose sanctions over allegedly unfair shipping practices at Japanese ports, they were largely manageable. The US continued to urge the Japanese not to try to export their way out of their economic troubles and to stimulate domestic demand instead, but clearly appreciated the depth of Japan's difficulties. Consequently, the main international issue of the year was the shaping of a new set of US–Japan defence cooperation guidelines, the first such revision since 1978 at the height of the Soviet threat. The new guidelines aroused controversy both within Japan and amongst Asian neighbours.

The reaffirmation of the Japan–US security alliance in April 1996 had allowed the two sides to begin serious reconsideration of the guidelines for cooperation in emergencies. The US, which had been disturbed by the weak attitude taken by Japan during the North Korean nuclear weapon crisis in the spring of 1994, pushed hard for this review. An interim report was issued in June 1997, followed by the final agreement in September. Taking into account the changes in geopolitics since the demise of the Soviet Union, it listed the many fields of cooperation, including the supply of materials (other than weapons and ammunition) by Japan to US forces, intelligence gathering, mine-sweeping, inspection of suspicious vessels during economic sanctions and evacuation of non-combatants and refugees. Both sides also stated that such logistical support would be provided by Japan not only within Japanese territory, but also in areas surrounding Japan.

Two aspects of the negotiations proved controversial: the exact nature of the new cooperation and, most crucially, the exact geographical area covered. In both ambiguity continued to exist. The US was happy to have at least a clearer definition of what Japan could be expected to do in an emergency, but some points were left unclear. For example, Japan could not 'supply' weapons, but could 'transport' them to US forces, provided that the transport remained outside the battle zone.

Because of clear differences of opinion, even within the LDP, the Japanese government found it hard to explain to its neighbours where exactly in the environs of Japan the new guidelines would apply. Defence Agency officials admitted that the scope of the new guidelines would be 'a little wider' than that encompassed by the term 'Far East' in the original Japan–US security treaty (a term taken to cover the area north of the Philippines). However, soon after the interim report was issued, the dovish Kato told the Chinese during a visit to Beijing that the guidelines were not framed with China in mind, but only North Korea.

The Chinese, however, remained suspicious that the guidelines were not just about a crisis on the Korean peninsula, but also applied to the Taiwan Straits. When hawkish Cabinet Chief Secretary Seiroku Kajiyama argued in mid-August that Japan could not refuse a US request for logistical support in the event of a Taiwan–China military conflict, Beijing became particularly incensed. In response, the Japanese Foreign Ministry continued to maintain that the Taiwan Straits were excluded. Consequently, when the final version of the guidelines was issued, attempts were made to smooth over the differences of interpretation by describing the concept as 'situational' not geographical. During a visit to China in September, and in a Tokyo meeting with then Chinese Premier Li Peng in November, Hashimoto carefully avoided giving a precise geographic definition. Whatever the Foreign Ministry may say publicly, in practice Kajiyama's interpretation is much closer to how the US interprets the geographical scope of the new Japanese commitment.

Before the final guidelines were issued, however, Japan found itself involved in a crisis which raised questions about the role of the Self-Defence Forces (SDF) outside Japan's immediate territory. In early July 1997, a Japanese engineer was killed when he was caught in the cross-fire between rival forces in Cambodia. On 12 July, three C-130 Hercules transport planes with 70 Air SDF on board flew into U-Taphao base in Thailand to stand by on alert to fly into Cambodia to evacuate Japanese nationals. After four days, during which the situation in Cambodia stabilised and those Japanese who wished to leave were brought out by other means, the aircraft were ordered back to Japan. This was the first time that the SDF had been sent overseas for an evacuation operation since the 1994 revision of the SDF Law to allow such missions. Hashimoto claimed that this action was taken to be prepared for the possibility of a greater emergency in Cambodia, while Doi Takako, head of the SDPJ, questioned the legality of sending the aircraft, as the revised SDF Law does not recognise such anticipation. Regardless of the humanitarian aspect (several other countries also sent military transport to evacuate their nationals), Hashimoto probably acted quickly on this occasion both to display his own decisiveness and to demonstrate to the US that Japan could take action if necessary in an emergency.

While the controversy over the guidelines rumbled on, so did the issue of the presence of US troops in Japan, specifically in Okinawa. US Secretary of State Madeleine Albright, on her first visit to Japan in February 1997, told Hashimoto that the presence of US troops in Japan was 'indispensable' to regional security. Hashimoto agreed with her, but added a request that the US consider cuts in the huge concentration of troops in Okinawa (see map, p. 243). Hashimoto was still being forced to deal with Okinawan resentment at the US presence there. Japan and the US had agreed in 1996 that the large US Marine base at Futenma in Okinawa should be closed down and be replaced by a heliport. Hashimoto struck a deal with the NFP in April to pass legislation allowing that, if individual landlords refused to renew leases on the land used by the US military, other land could be used for the bases instead. An inter-ministry task force on Okinawa, headed by Defence Director-General Fumio Kyuma, then tried to seduce the Okinawan electorate by linking the relocation of Futenma and the creation of the heliport to economic development plans. The task force hinted these could even include turning Okinawa into a free-trade zone (an idea advocated by the Okinawan governor). This did not impress the inhabitants of the city of Nago, the proposed site of the heliport, who voted against it coming to their neighbourhood in a referendum in December 1997. The Okinawa issue was set to remain a problem for Hashimoto throughout 1998.

Thaws In North-east Asia

Unlike the Americans and the Europeans, the Japanese were sceptical that the end of the Cold War and the collapse of the Soviet Union would mean much change in their difficult relationship with their northern neighbour. Even with the arrival of Boris Yeltsin's Russia, Russian troops still occupied the four 'northern territories' claimed by Japan, significant armed forces were still stationed around the Russian Far East, and Russia had neither the means nor the will to make its still ramshackle economy really attractive to hard-headed Japanese businessmen, despite its undoubted natural resources.

From early 1997, however, there were signs of quiet moves to take the chill out of the relationship. And it was the Japanese, previously so cool, that were pushing forward. Ironically, a maritime accident provided the impetus. Russian cooperation in investigating the sinking of a Russian oil tanker in the Sea of Japan in January 1997 impressed the Japanese, who had been used to obstructionist Soviet tactics over past maritime incidents. Hashimoto decided to take the opportunity to make a gesture to the Russians. In March he told US President Bill Clinton that he had no objection to Russia formally joining the Denver summit discussions, thus enlarging the G-7 Group of leading industrial nations to G-8. Although many Japanese officials still felt that Russia's economic situation did not really qualify it to participate, they recognised the country's strategic importance. Yeltsin appreciated the Japanese gesture. At his meeting with Hashimoto in the margins of the Denver summit in June he told the Prime Minister that Russian missiles were no longer targeted on Japan, that he had no objection to the revamping of the Japan–US security treaty and that he supported Japan's bid to become a permanent member of the UN Security Council.

The two leaders also discussed ways to develop a security dialogue. Yeltsin was following up on a visit to Tokyo in May by his then Defence Minister, General Igor Rodionov, who had surprised Japanese defence officials by offering to let them visit any military bases they wanted to in Russia and by proposing exchanges of Russian, Japanese and US defence personnel. He also added that cutting down the Russian forces was expensive, so any financial assistance from Japan would be welcome .

The major issue between the two countries remains unresolved. Although diplomatic relations between the Soviet Union and Japan were resumed in 1956, no peace treaty ending the Second War War was ever signed. Negotiations since have been intermittent and unsuccessful. The Soviet Union under Mikhail Gorbachev, and subsequently Russia, occasionally hinted that some form of concession over the territorial issue might be available if the Japanese made the first move with an offer of economic aid and investment. The Japanese had always refused to move.

During 1997, however, the Japanese gave signs that they might be willing to develop economic relations without waiting for the territorial negotiations to make real progress. Russian companies still owe $1.1bn to Japanese companies. On a visit to Tokyo in June, the new reformist Russian First Deputy Prime Minister Boris Nemtsov promised that his government would make serious efforts to solve the debt problem. Japanese government officials and business organisations were impressed by Nemtsov's apparent determination to reform the Russian economy, and Japan's Import–Export Bank made three loans totalling $500m to help economic reform measures and encourage Japanese business participation.

The territorial dispute remains tricky, but Hashimoto told Japanese businessmen that 'we cannot resolve the dispute if it results in one side being the winner and the other a loser' – the softest-ever stance on the question by a Japanese prime minister. The informal summit between Hashimoto and Yeltsin in Krasnoyarsk in November 1997 set the seal on this new relationship. In what was almost certainly the warmest personal meeting between Russian and Japanese leaders this century, the two agreed to work towards a peace treaty by the year 2000 and to enhance economic contacts.

The regional balance is clearly an issue in the emerging relationship. Both sides share a worry about the more nationalistic and militarily-powerful China that is emerging. The Russians and the Chinese have improved their relations noticeably since early 1997, but, as the Russian Defence Minister General Rodionov privately told his Japanese hosts, there are limits to how far the Russians can expect that relationship to develop.

While Japan is careful to avoid specifically mentioning China as a potential threat in any of its defence policy statements, it has also become increasingly concerned about the Chinese defence build-up. Japan's relations with China did improve during 1997, after what Hashimoto himself described as a 'somewhat awkward' period. The territorial dispute over the Senkaku/Diaoyu islands simmered, fuelled by a landing on the disputed islands in May 1997 by an NFP politician. By the end of the year the two sides had come to a new fisheries agreement which effectively put the territorial dispute back on the shelf. However, activists in Japan, Taiwan and Hong Kong may not be content to leave it there.

Hashimoto, on his visit to China in September 1997 to celebrate the twenty-fifth anniversary of diplomatic relations, brought a gift of new yen loans for environmental projects and spoke of his distress on visiting sites of past Japanese atrocities in north-east China. Neither the rhetoric surrounding Hashimoto's visit and Premier Li's return visit to Japan two months later, nor the positive atmosphere engendered by the first-ever visit of a Chinese Defence Minister to Japan in early February 1998, could wholly obscure the irritable tone which has crept into Chinese media commentaries on Japan, and the caution with which many Japanese officials and politicians still view China in private.

Japan also remained nervous about famine-wracked North Korea, but here too there were signs of a slight thaw in relations, as North Korea finally agreed to allow a few elderly Japanese wives of North Koreans to return to their homeland for brief visits (though implicitly only in return for Japanese food aid) and the two sides edged towards re-opening the negotiations for establishing diplomatic relations which were suspended in 1992. Nevertheless, the South Korean economy is under strain and its government has been suggesting that the heavy payments due under the Kedo agreement for the construction of nuclear plants in North Korea might need to be re-negotiated at a time when Japan itself is having economic difficulties. This conjunction of events will ensure that the triangular relationship will remain a tricky one for Japan to handle.

Self-destructing Opposition

Towards the end of 1997, Hashimoto's popularity was waning. He began the new year with public opinion polls showing support for him at around the 30% level, the lowest since he became prime minister. He tried to regain credibility by pushing ahead with his reform programme. In early December, the cabinet finally agreed to a far-reaching reorganisation of the administrative structure, based on the final report of an ad hoc administrative reform panel chaired by Hashimoto himself. Designed to streamline the bureaucracy and slash the number of ministries and agencies, the blueprint calls for the Prime Minister's Office to be revamped into a new more powerful Cabinet Office. The current total of 21 ministries and agencies will be consolidated into just 12 ministries. The various economy-related ministries and agencies would be combined in ways which would reduce the independence of politically-powerful ministries such as the Construction Ministry and the Ministry of International Trade and Industry. The Foreign, Justice and Agriculture Ministries would survive in virtually their present form, while the Environment Agency would be upgraded into a ministry. The Defence Agency, despite heavy lobbying from the right-wing of the LDP for an enhanced status, will not be changed, primarily because of SDPJ objections. The most controversial aspect of the reform is the reorganisation of the powerful Ministry of Finance (MOF), which is to be renamed the Ministry of the Treasury. Disagreements between the LDP and its allies forced the cabinet to postpone a decision on whether the MOF's budgetary powers would be separated from its authority to supervise financial institutions (crucial, given the number of financial scandals in Japan).

Heralded by government spokesmen as the most ambitious reform of the Japanese administrative structure for more than a century, the real test will come in drafting and trying to implement the precise legislation. The Japanese business community generally welcomed the plans, but the opposition NFP labelled them a sham, arguing that just renaming ministries was not real reform. The most effective opposition, however, is expected to come from the bureaucrats themselves, who will undoubtedly wish to dilute the reform's impact on their own particular agencies and ministries. The fact that the LDP has to pass two bills, a basic one in the summer of 1998 on the principle of reorganisation and a more detailed one in 1999 on the precise details of the new ministries, gives plenty of opportunity for bureaucratic obstruction and parliamentary squabbling. The start of the new structure, set for January 2001, is likely to seem a long while coming.

Meanwhile, Hashimoto has been handed an unintended life-line by Ozawa, the opponent he earlier 'rescued'. Despite criticisms from within the NFP, Ozawa did nothing to change his abrasive management style. Increased intra-party tensions and the continuing trickle of defectors provided anti-Ozawa groups within the NFP with their opportunity and they put up Michihiko Kano as a rival candidate in the party presidential elections in December 1997. Ozawa won comfortably, but in the process sealed the fate of the NFP. Within a couple of weeks it had split into six new parties. Ozawa's mainstream group, renamed the Liberal Party, was left with only 42 members in the Lower House. Several of the new parties expressed an interest in working together with the DP, which has now become the focus of moves to create a new large opposition party, provisionally entitled the *Minyuren*, which would have around 100 members.

With the opposition in such disarray, Hashimoto can now concentrate on the big decisions, particularly on the economy. Whether history records him as the only Japanese prime minister of the 1990s with any vision or as yet another run-of-the-mill LDP 'manager' depends on how he tackles them in the coming year.

◆

The Korean Peninsula: A Year Of Surprises

At the beginning of 1997, the North Korean economy was widely predicted to be on the verge of collapse, and that therefore the North Korean regime was doomed. The year ended with the situation still unresolved. Once again, natural disasters dealt agricultural output a body blow, and, once again, the North appealed for international assistance to feed its population. The aid was forthcoming and the North staggered through the year, but the prospect still looms of, at best, continued food shortages, and, at worst, famine. Somehow, despite occasional defections and the apparent destruction of the country's industrial base, the political structure remained intact. Although Hang Jang Yop, one of the country's senior ideologues, defected in February 1997 this did not signal a major split in the ruling Korean Workers' Party. By the autumn, Kim Jong Il was confident enough to accept the post of General-Secretary of the party, one of the two posts (the other was head of the government) that his father, Kim Il Sung, had occupied after he had consolidated his power.

South Korea produced even more surprises. There had been signs of economic problems in 1996, and the Hanbo iron and steel company had declared bankruptcy early in 1997. Further industrial failures followed later in the year, but few suspected the real extent of the problems which 'Korea Inc.' faced until the storm hit in November. By the end of the year, bankruptcies were coming thick and fast, the *won* had lost half its value in under six weeks, and the once proud Korean economic miracle was being shored up by the largest support package ever from the International Monetary Fund (IMF), a total of $57 billion. Yet the incumbent president, Kim Young Sam, failed to provide leadership, as the country prepared to elect his successor. In a complete break with the past, and affirming the growing maturity of South Korea's democracy, the voters turned against the ruling party, and, for the first time in an election that also featured a candidate from the ruling party, elected a leader of the opposition, Kim Dae Jung to the presidency.

Despite the long crisis in the North and the economic upheaval in the South, there was no major change in the security situation on the peninsula. If anything, there was some small progress towards greater stability in the relationship. The two Koreas, China and the US finally got around a table in Geneva in December 1997, and the Four Party talks, proposed by Presidents Bill Clinton and Kim Young Sam in 1996 as a possible means of reaching a peace settlement, got under way. In February 1998, just before Kim Dae Jung (who had made improving diplomatic ties with the North a top priority) was to be sworn in as the new president of South Korea, North Korea made an unusual request for a thaw in the frozen diplomatic relations between the two countries. It sent some 70 letters through the Red Cross to various officials and bodies in South Korea calling for negotiations and political dialogue, suggesting that a significant easing of tensions was now not only possible, but indeed was underway. Another positive sign was that, despite occasional difficulties, the 1994 nuclear deal between the US and North Korea held up, and the Korean Peninsula Energy Development Organisation (KEDO) was able to begin work under the agreement on the Kumho site, near the coastal city of Sinpo in North Korea.

The North: Kim Goes Up, But The Economy Goes Down, Down

The defection of the North's top ideologue, Hwang Jang Yop, in Beijing in February 1997, was followed by a period of bargaining with the Chinese to move him safely out of China. Once out, and established in Seoul, he was eventually allowed to go public. Much of what he said turned out to be predictable, and it was clear that he had not been in the inner circle of North Korean politics for some time. He was, nevertheless, able to fill in some of the details of Kim Jong Il's style of work and how the system functioned.

Some saw his defection, and the defection in July of the North Korean ambassador to Egypt and his brother, as further signs of the imminent collapse of the North's political structure, but this proved unfounded. These defections, serious as they were individually, did not foreshadow a significant flow of senior officials from the North. The North even got one prize going the other way when a South Korean religious leader with links to the opposition National Council for New Politics fled north in August. There were a few awkward moments for the opposition in Seoul, but they were soon over. In much the same way, the embarrassment caused by the high level defections from North Korea soon passed, and the shaky regime continued to survive.

One source of speculation and rumour was removed in October 1997 when Kim Jong Il moved to take up one of the posts that his father had held; he became Secretary-General of the Korean Workers Party. It was an indication, however, of the unsettled state of the North, that this appointment came about by 'acclamation' at a series of local party meetings, rather than in accordance with the party constitution. At the same time, the media frequently indicated that Kim Il Sung was still, in some way, present among his people, and it regularly repeated the mantra 'Kim Jong Il is Kim Il Sung'.

Kim's assumption of the top party post did not foreshadow any obvious change of policies. He still seemed to prefer his military title and to identify with the military. This has often been interpreted as evidence of his dependence on the armed forces, though the close ties between the military and the top leadership in North Korea, as in China, date back to the days of guerrilla warfare in the 1930s.

Accounts of North Korea's food problems appeared regularly in the world's press, and at one point it was widely predicted that a major crisis would occur by June as the food from the 1996 harvest ran out. The international aid agencies made a series of appeals and the international response kept North Korea going during the late spring and early summer, when the poor 1996 harvest had been exhausted. Whatever hopes there were for the 1997 harvest did not last long. The rice harvest was not as bad as in the previous two years, but the corn harvest was a disaster as drought withered crops in the fields. There was food, but by no means enough.

Again, the North Koreans appealed for aid, and again there was a strong response from the West, from a series of non-governmental organisations, from South Korea and from China. The total amount of such aid can only be guessed at because the Chinese preferred not to publicise what they were giving, but it was substantial. The United Nations Development Programme estimates that China provided some one million tons of food in aid and barter, and the North Koreans have indicated that they received some 600,000 tons from other sources. Although it helped, this aid did not meet the North's total shortfall. Even when all sources of food are taken into account, the North will need another 1,034,000 tonnes in 1998 according to South Korean government estimates, or 1,710,000 tonnes according to estimates by the Food and Agriculture Organisation and the World Food Programme. This may not be famine, but it is slow starvation. There is little margin for animal feed or for industrial production.

Economic life of a sort did continue, though the planned economy of the past seems to have been abandoned. There has been no central plan since the third seven-year plan ended in 1993 without achieving its goals. Local markets multiplied where those who had money could buy food. There were also reports of wholesale barter trade along the border with China; Chinese officials seem to encourage trade that at least keeps the North Koreans on their own side of the border. There was also evidence that some factories had begun to function during the spring and summer of 1997, though they closed down again as the cold weather set in with the autumn. A serious set-back was the lack of fuel due to the absence of hard currency to pay for it; this has affected the ability of the regime to keep factories running and to provide transportation for the distribution of food and other goods.

An IMF team, which visited North Korea in early September, estimated that industrial production had declined by at least 60% in five years. The crisis was also spreading to areas that had hitherto been protected, with reports from Pyongyang itself that water and electricity were only available on selected days. More serious for the long-term future of the economy were reports that factories were selling off machinery to China as scrap in order to raise immediate funds for food, and one defector indicated that the North Korean transportation system had virtually broken down. Against these accounts, the news that North–South trade had reached $3 million seemed a minor matter. Yet even this figure may not survive the economic downturn that the South is now undergoing.

South Korea: Miracles No More?

Hints that all was not well in South Korea were given when the Hanbo iron and steel complex collapsed early in 1997. There did not immediately appear to be any wider economic implications, although Sammi Steel followed Hanbo in March. Both enterprises had been overextended and, perhaps, over indulged by a complacent central government, but 'Korea Inc.' seemed sturdy enough to ride the storm. Nevertheless, there were undoubted political repercussions. Persistent claims of wrongdoing on the part of Hanbo executives eventually led to the Chief Executive and others from the top layer of the company being indicted, and ultimately convicted, on corruption charges. To make matters worse, corruption charges reached out to one of President Kim's sons, who was arrested in May, accused of bribery and tax evasion.

Kim seemed untouched, though there were those who wondered on whose behalf the younger Kim had engaged in the actions for which he was eventually convicted. Kim Young Sam's main preoccupation, however, was the presidential election, due in December. He made it clear that he did not want his old rival, Kim Dae Jung, to win. The ruling party's argument was that the era of the 'three Kims' – Kim Young Sam, Kim Dae Jung and Kim Jong Pil – who had dominated the political scene since the 1970s – was over. Now younger, less tainted politicians should take over.

This was all very well for Kim Young Sam who had become president in 1992 in what the other two Kims regarded as a shady deal with the then president, Roh Tae Woo, and who was barred by the constitution from seeking a second term, but the other Kim's needs were different. Not surprisingly, they refused to be compliant and both indicated their determination to fight. Their task seemed difficult at first as the President's ruling New Korea Party put forward a respected High Court judge and former prime minister, Lee Hoi Chang, as its candidate.

Lee had all the appearance of the required break with the past. He was somewhat younger than the two Kims and had a reputation as an honest man. He soon seemed more a liability than an asset, however. This was partly the result of tensions within the ruling party itself, as many who had rallied around a possible alternative and younger candidate, Rhee In Jae, the Governor of Kyonggi province, indicated that they did not wish to support the official candidate and might form a breakaway party. More damaging, however, was the revelation that Lee also had some blemishes on his 'Mr Clean' image. In particular, both his sons had apparently evaded military service. Electioneering, although officially not allowed until November 1997, was well under way by the early autumn, with Kim Dae Jung taking an early, but by no means conclusive, lead in the public opinion polls.

It was at this point that the economic roof began to fall in. In July, Kia Motors, the country's third-largest automobile manufacturer, had sought government assistance as it could not meet its short-term loan requirements. By October it was bankrupt and the government stepped in to nationalise it when the banks refused further loans. The government was apparently fearful of the social consequences of not taking action when there was virtually no social security system available as a safety net for the work force. International credit rating agencies began to mark South Korea sharply down. By mid-November, the exchange rate of the South Korean won had dropped below 1,000 to the US dollar. The government denied it needed assistance and the Bank of Korea spent billions of dollars trying to protect the won. It failed and, on 21 November, the government approached the IMF for $20bn to help meet short-term debt. A week later, the IMF granted South Korea $57bn, the largest sum ever provided. At that stage, South Korean external debt was estimated at $156.9bn.

The immediate response in South Korea was despair. The media were full of images of Koreans protesting at the IMF 'humiliation'. Students were shown jumping on Japanese pens, and there were loud calls for a halt to foreign travel and an end to the purchase of foreign goods. The IMF was accused of following the US line in insisting that the South Korean economy be opened up to outside competition. Despite the pain felt by South Korean consumers, the bottom had not been reached. On 22 December, the international credit agencies reduced South Korean state and corporate bonds to junk bond status, and the won fell to just under 2,000 to the dollar. It bounced back, going into the new year at about 1,800 to the dollar, and later reached 1,500 to the dollar, but the pill was a bitter one.

Government leadership was most notable for its absence. Rumours abounded that President Kim Young Sam had been virtually unaware of the crisis until mid-November; he was clearly uncertain what to do thereafter. Nobody could be sure what effect the economic crisis would have on the election. Forecasts of the result were complicated by splits in the ruling party, the emergence of new candidates, and the difficulty of judging the impact of the very changed circumstances in which the Republic of Korea now found itself.

In the end, it was Kim Dae Jung who won on 18 December, but only by some 400,000 votes out of 26m, and in an unlikely alliance with Kim Jong Pil, the long-lived architect of the 1961 military coup. The economic crisis overshadowed Kim's success, but it was of great significance in South Korea's journey down the road to democracy. For the first time, the electorate had returned a candidate who did not have the full backing of the state behind him. It might have been a victory by the narrowest of margins, but victory it was nevertheless.

How Kim Dae Jung would handle the collapsing economy was the most immediate and dominating question. Although he had called for economic reforms in the past, he was generally regarded as a populist and he certainly had no experience of government. In his first days as president-elect, he seemed to make matters worse by describing his shock as official briefings revealed how bad things were. He later claimed that this was a deliberate ploy to bring home to the Korean people the seriousness of the problem, but not everybody was convinced. Before long, however, he began

to use his prestige to persuade both trade unions and the giant conglomerates of the need to accept the stringent reforms the IMF had laid down as a
requirement for its loans. He had also begun to assemble a strong team of
advisers, while at the same time working closely with Kim Young Sam and
the outgoing administration. One of the first things on which they reached
agreement was the release of former presidents Chun Doo Hwan and Roh
Tae Woo, who had both been convicted on mutiny, treason and corruption
charges. They left jail on 22 December, but they are still supposed to pay the
massive fines that were originally imposed.

Talking At Last

For the most part North Korean and South Korean relations appeared stuck
in a time warp. The North repeated its long standing proposals for a unified
Korea and continued to call for a peace treaty with the United States to
replace the 1953 Armistice Agreement. There was a serious incident in the
Demilitarised Zone in July, when North Korean troops crossed into the
southern half of the zone, while North Korean guard posts exchanged fire
with South Korean posts. In response to protests by the UN Command, the
North denied that it had done anything, but the incident was not repeated.
There was a certain ruffling of Southern feathers when the North threatened
to blow up the South's state-owned broadcasting company and two
newspapers for their anti-North Korean stand, but nothing came of the
threats. A major Northern spy-ring was exposed in the autumn. The
South's economic problems caused a certain amount of glee in the North,
though this was tempered, perhaps, by the realisation that there could be
adverse consequences for the North itself. There was some gloating over the
failure of Kim Young Sam's chosen successor to win the presidential
nomination, but the North seemed unsure of what to make of Kim Dae
Jung's election and did not report it for some time. In the South, the new
president was preoccupied by the economic crisis, and he indicated that
policy towards the North would not be his highest priority.

There were some changes, however. The North called for and received
South Korean food aid, and acknowledged that it had done so. It also
attended a series of preparatory meetings designed to pave the way for the
four party talks proposed by Presidents Clinton and Kim Young Sam in
April 1996, and finally attended the first plenary session held in Geneva on
9–10 December. This meeting did little more than agree to meet again in
Beijing in March 1998, but that in itself was an achievement. The initiative
taken by North Korea in February 1998 to open a wide-ranging dialogue
with South Korean officials and organisations boded well for the March
meeting. In the event, however, the North continued to throw up obstacles
during the talks, and although they were still underway on 26 March there
were not yet signs that significant advances would be made.

Meanwhile, the Korean Peninsula Energy Development Organisation (KEDO), started work in North Korea. KEDO is the international consortium formed to implement the key elements of the 1994 US–North Korean agreed framework, under which Pyongyang has frozen its nuclear programme in exchange for the construction of two light-water reactors. Work preparing the site for the reactors at Kumho, near Sinpo, began in August 1997. This was made possible by assurances given by the International Atomic Energy Agency (IAEA) that the North Korean nuclear complex in Yongbyon remained shut down under its monitoring and that some 75% of the spent fuel from North Korea's plutonium-producing gas graphite reactor had been packaged, or canned, for shipment out of North Korea, as required by the agreed framework.

In December 1996 a supply agreement was concluded which established the scope of supply for the two light-water reactors; set terms for repayment by North Korea of the cost of the reactors; and set the general terms and conditions under which KEDO operates in the North (covering transportation, communications, and privileges and immunities for KEDO and the employees of the contractors). A large number of protocols to this agreement had to be worked out before site work could begin, but, although those presented some difficulties and problems, they were all solvable. The expert-level delegation that held most of the talks in Sinpo even established a new precedent, travelling to North Korea directly by ship from the South rather than flying in from a third country as Pyongyang originally wanted.

There are still a number of difficulties to be overcome. One of them concerns financing the project. The financial crisis in South Korea has raised the question of whether South Korea is still prepared, and able, to pay the bulk of the $5.18bn price tag which it initially assumed. In early January 1998, President-elect Kim Dae Jung argued that the financial crisis meant that 'Japan and the United States must contribute more'. While this will clearly be the subject of continuing debate, the United States does not appear to be in much doubt that South Korea and Japan will be able to continue with KEDO's reactor construction.

Subsiding Tensions

In their different ways, the two Koreas face an uncertain future. That of the North is the more gloomy. Its economic decline shows no sign of coming to an end. The regime has made some attempts at change and seems to be prepared to tolerate more. It has established a number of free trade zones, mostly on the coast, and by the beginning of 1997 at least $333m had been invested there with over 300 joint ventures starting up. However, there is as yet no evidence of a will to tackle the fundamental organisational and structural problems which are at the heart of its difficulties. It is these which have caused the breakdown of both agriculture and industry, exacerbated though the situation may have been by natural disasters.

The South's short term prospects are grim. As the economic reforms imposed by the IMF begin to bite, bankruptcies among smaller businesses, already averaging over 100 a day by January 1998, are likely to increase with distressing consequences for the owners and all they employ. A new government, many of whose members face a steep learning curve as they discover the difference between opposition and government, will have new and unfamiliar problems to face. One will be the relationship between the new president and the National Assembly, which will be controlled by a hostile party. It may well be, however, that the Grand National Party, as Kim Young Sam's party was titled in the course of 1997, will begin to break up as its members discover the disadvantages of no longer having their leader in the Blue House.

Another likely prospect is growing antagonism towards foreign influence as the consequences of the forced opening of South Korean markets to overseas competition begins to be felt. Like their counterparts in the North, South Koreans tend to discount the structural weaknesses which have created their problems, preferring to blame their plight on unfriendly forces abroad. Kim Dae Jung will face a tough task in persuading them to address the real issues. That said, much of what made South Korea successful in the first place is still there and, while the next year or two will be difficult, the country can be expected to pull through this crisis in the end. The crisis may even do some good for it provides a unique opportunity to re-structure the economy in areas where reform would not have been possible in normal times. By early 1998 there was already more than talk underway about breaking up the *chaebol* (South Korea's huge conglomerates) and severing their links with the government.

What all this means in terms of security is not yet clear. The South's economic difficulties may limit both its willingness and its ability to help the North as the latter's food crisis enters its fourth year. At the same time, the South, which is already wary about reunification because of the cost, is likely to be even less enthusiastic as its economy contracts. Suggestions that the South might not be able to meet the large share of the costs of KEDO which it has agreed are a particular concern. KEDO is an essential part of the 1994 agreement designed to stop the North developing nuclear weapons, and therefore essential to the continued security of the peninsula.

The economic woes of North and South may be beneficial in the security context, though it would be a mistake to be too complacent. North Korea still has the ability to inflict considerable damage on the South, especially since Seoul is so near the Demilitarised Zone. However, there are signs that the long-drawn out economic decline is beginning to affect North Korea's armed forces along with everybody else (although they are widely believed to get priority in food distribution). Training has been reduced, and shortages of fuel will undoubtedly affect operational capabilities. The punch is still there, but its effectiveness may be slipping away, and the continued need for international aid does provide a lever on the North that did not exist ten or even five years ago.

Economic difficulties may also force the South to rethink its policies on the peninsula, a process which may be helped by a new president. Already there have been references to the need to reduce defence expenditure, but, in due course, more fundamental reassessments of Seoul's policies towards the North are likely. Kim Dac Jung's willingness to open a serious dialogue with the North, coupled with the indications in early 1998 of a new desire on the part of the previously insular regime to continue the talks with the South, have created a more encouraging atmosphere. Perhaps the changes in both the North and South in 1997 may lead both into a more constructive relationship.

◆

India: Struggling With Political Deadlock

The uncertainties of party politics and government instability, together with the need to implement market reforms and maintain a sound economy, have been the primary concerns in India and the rest of South Asia during most of the 1990s. These newer concerns overshadow traditional internal and external security issues: the problems of secessionist violence and domestic religious, linguistic and caste conflicts. No satisfactory answers have been found to any of these problems.

Indian voters had another opportunity, in the elections of February and early March 1998, to create a strong government, led by a majority party determined to carry through its manifesto. Instead, as had been the case only two years before, they produced another hung parliament, with the major parties furiously bargaining in an effort to form a minority government. The resultant weak coalition, headed by the *Bharatiya Janata* Party (BJP), faces the constant threat of dissolution and is thus little better placed to overcome India's malaise than the four governments that preceded it.

Shaky Governments In India

Weak coalition and/or minority governments have ruled India throughout the 1990s. In 1991, V. P. Singh's National Front coalition, which came together in 1989, collapsed. The minority Congress government then formed by P. V. Narasimha Rao had to rely on support from parties on the ideological left on certain issues, and the BJP, on the religious Hindu right, on others. Although it barely weathered one no-confidence vote, it managed to survive for almost its full five-year term.

The secular United Front (UF) governments of prime ministers H. D. Deve Gowda and I. K. Gujral were not so fortunate. After the 1996 elections, the UF had been able to cobble together from several diverse parties and independent groups only 183 seats in the 545-seat *Lok Sabha*, the lower house of India's bicameral legislature. By comparison, the BJP held 162 seats and could count on the support of another 31 members from the parties of its allies, including the even more radical Hindu nationalist party, *Shiv Sena*. Since the BJP would provide no support for the strongly secular UF, the Congress party, with only 140 seats, became the main pillar propping up the UF.

In 1997, the Congress party twice withdrew its support for the UF government leaving it without a majority in parliament. Sitaram Kesri, the ageing but still ambitious leader of Congress, first withdrew support in April in the hope of forging a minority Congress government with himself as prime minister. When this effort failed, he offered to renew support for the UF if it dropped Deve Gowda and reconstituted itself under a different leader. The UF tamely obliged, although there appeared to be nothing wrong with Deve Gowda's performance. When Kesri's Congress party withdrew support a second time in December, Prime Minister Gujral of the UF could no longer muster a majority in the *Lok Sabha*. This led to the fall of yet another Indian government, and yet another round of national elections in February–March 1998. The intense jockeying for position that followed the polls, and the personal ambitions of politicians to become prime minister, or a minister of any sort, whether meaningful or not, may be indicative of future Indian politics, particularly since neither the Congress nor the BJP were able to win a clear majority. The BJP attained its highest-ever total of votes and seats, but even with the support of its election allies, fell at least 20 seats short of the 273 needed for a majority. Congress and its ally the UF, were at least ten seats short of a majority.

After two weeks of complex discussions between the two leading parties and their possible allies, it became clear that Congress would be unable to secure enough seats to form a government. President Kocheril Raman Narayan turned to the BJP, which attracted sufficient support from minor parties to begin forming an administration.

The BJP Thrives

Until the election campaign of early 1998, the BJP and other, more radical, Hindu parties such as *Shiv Sena*, the *Vishwa Hindu Parishad* (VHP) and the *Rashtriya Swayamsevak* (RSS), argued for the rejection of the secular ideals of Congress, and, indeed, of the secular democracy espoused by most other Indian parties. Hindu nationalists wish to establish India as a Hindu state to be called *Hindutva*. The benign interpretation of *Hindutva* would be a broad Hindu world view within which other religious groups would take part; the VHP and RSS, and the more extreme elements of the BJP and *Shiv Sena* wish, however, to marginalise the minorities in India.

In early 1998, with the BJP on the threshold of gaining even more seats in the *Lok Sabha*, the party moved closer to the secular ideology of the other parties. The BJP's main leader and designated shadow Prime Minister, A. B. Vajpayee, publicly declared in January 1998 that a BJP government would retain India's secular constitution and would continue to implement the secular policies of previous governments. Such reassurances were reinforced by similar statements from BJP President L. K. Advani (who was appointed Home Minister in the cabinet formed by Vajpayee in March).

There are at least two reasons for this ideological shift on the part of the BJP and the *Shiv Sena*. Unlike the earlier Hindu *Mahasabha* party, which was banned after one of its members assassinated Mahatma Gandhi, and its successor, the *Jana Sangha*, from which the BJP originated, the BJP has learned that an avowedly Hindu nationalist platform would create fear and provoke adverse reactions among India's 120m-strong Muslim minority, as well as among the country's 50m Christians and Sikhs. Secessionist violence in Kashmir, Punjab and the Christian tribal states of the north-east could be aggravated.

Additionally, the emphasis on Hindu nationalism, with its appearance of Hindi-speaking northern domination, has not generated support from the Dravidian states in the south – Tamil Nadu, Andhra Pradesh, Karnataka and Kerala – which contain about 20% of India's population. More importantly, the image of Hindu nationalism represented by upper-caste Hindus has alienated the *Dalits* (the former 'Untouchables') and the other 'Backward Castes' (the lower-caste *Shudras*), which together constitute about 70% of the Hindu population and about 55% of the Indian population as a whole. Without their support, the BJP would have difficulty obtaining an electoral majority in the *Lok Sabha*, and it would return to being the marginal party it was for the decades before the 1990s. Democracy in India has, in fact, reversed the power of the Hindu caste hierarchy, with upper caste Brahmins accounting for the smallest number of votes, probably less than 5%.

Based on its proclaimed shift from Hindu nationalism to secularism and regionalism, the BJP forged electoral alliances with various regional parties. It is not clear whether this new and improved image represents a permanent change, or whether it is merely a temporary electoral strategy. Vajpayee's stated commitment to secularism was often countered by Hindu nationalist statements from other members of the BJP, such as Murli Manohar Joshi and especially Ashok Singhal of the VHP, both of whom reiterated their commitment to a Hindu state. But the need to rely on minor, non-Hindi, parties to stay in government will keep the BJP from straying back to its more extreme positions.

In an early indication of the more moderate stance that it realises it must take to satisfy the 14 small parties that go to make up the coalition government, the new BJP administration immediately issued a declaration on a 'national agenda for governance'. This omitted virtually every stance previously held by the BJP which had caused anxiety among the non-Hindu population. Vajpayee, who was sworn in as prime minister on 19 March 1998, noted in his press conference that the new government would not even attempt to put into practice its long-standing position that nuclear weapons should immediately become part of India's military arsenal. Instead, he fell back on the position held by almost every earlier Indian government: the country would 'keep its options open' with regard to its undeclared nuclear weapons status.

The Congress Party Fades

Congress has continued to suffer from the lack of strong leadership resulting from the assassination of former Prime Minister Rajiv Gandhi in May 1991 by a Tamil suicide killer. Former prime minister Narasimha Rao is embroiled in corruption charges and the political antics and tactics of the party leader, Sitaram Kesri, who toppled two governments in the vain hope of coming to power himself, eroded support for the party. The disarray and desperation that prevailed in its ranks in early 1998 drew Rajiv's Italian-born widow, Sonia Gandhi, and their daughter, Priyanka Gandhi Vadra, into the election campaign. Sonia Gandhi was reluctant to become involved, but the participation of the two was vital to the improved electoral fortunes of Congress. Although they were not able to turn the tide completely, their efforts on the hustings did ensure that Congress would do no worse than it had at the last election. Given the direction in which the party was heading before they took part, that was a signal triumph. After the election, Sonia Ghandi agreed to take up the position of president of the party, vacated under protest by Kesri, thus indicating that she was prepared to play a greater role in its fortunes.

It is easy to forget that Congress never achieved an overall majority of votes in any of its electoral victories, whether under Pandit Nehru or Indira and Rajiv Gandhi. Indeed, no party in India has ever done so, no matter how many seats it secured in the central parliament. Even when Rajiv Gandhi's Congress party won 80% of the seats in parliament in 1984, it received only 48% of the vote. In 1977, it was the banding together of most opposition parties that brought down Indira Gandhi's government. Unfortunately, convinced that it can ride to success on its historic name and the fame of its founders, Congress has made less serious efforts to forge electoral alliances than its rival, the BJP. Nevertheless, the secular-oriented Congress and the UF group of parties remain determined to thwart the designs of a Hindu nationalist BJP government. This 'negative bonding' will ensure another period of weak government under the BJP.

Economic Performance And Stability

Despite the political weaknesses that have bedevilled successive governments in the 1990s, they have been able to put into practice an economic liberalisation programme that has paid high dividends. In 1991–92, when liberalisation began, gross domestic product (GDP) was growing by 0.4% a year. The growth rate climbed steadily to reach 6.8% in 1995–96, and is expected to be 6.6% in 1997–98 according to projections by the International Monetary Fund. In the same period, external trade and foreign investments in India have increased several-fold. Between 1991 and 1996, the value of foreign investment approvals rose by $35bn. Foreign direct investment during this time amounted to Rs 150bn.

Table I India: Economic Indicators, 1991–1997

	1991	1993	1995	1996	1997
Average annual exchange rate (Rupee/US$)	22.72	30.49	32.43	35.43	36.31
GDP (US$bn)	258	285	339	371	405
Real GDP growth (%)	0.4	4.8	7.3	6.8	6.6
Consumer Price Inflation (%)	13.5	6.4	10.3	8.9	7.1
Exports of goods and services (US$m)	23,020	27,123	38,014	33,054	19,804
Imports of goods and services (US$m)	27,032	30,605	48,225	37,375	20,292
Balance on goods and services (US$m)	4,012	-3,482	-10,211	-4,321	-488
Gross external debt (US$m)	85,516	93,968	93,766	94,334	114,000
Foreign direct investment (actual inflows in US$m)	155	574	1,517	3,437	n.a.

Notes   1 1997 figures refer to the Indian financial year from April 1997 to March 1998 and are projections.

2 Export and import data for 1996 and 1997 exclude services, and for 1997 are for first six months only.

Sources   IMF, World Bank, Indian Government

The average GDP growth-rate of 6–7% since reforms began in 1991–92 is more than double that between 1960 and 1980. This economic growth appears to be modest when compared with the more spectacular average growth rates of 8–12% in China, Indonesia, Malaysia and Thailand. Yet India may display greater economic depth and stability than many of the other Asian countries, if judged in the context of wider issues of economic and political development strategy. The collapse, in late 1997, of currencies and financial markets in Thailand, Malaysia, Indonesia and South Korea with their varying degrees of authoritarian government and state-supported, export-oriented economies, underlined the stronger base on which reforms are being pursued in India.

Nevertheless, the Indian economy was partially affected by the economic crisis in East Asia. This was reflected mainly in the drop in the international market value of the rupee, and a drop in India's exports to the troubled Asian countries because of the very sharp fall in their currencies. By comparison, India's currency could be considered almost stable. From approximately 36 rupees to the dollar in mid–1997, the value had dropped to 39 rupees to the dollar at the end of 1997, and remained between 39 and 40 rupees in the first quarter of 1998. The major devaluation of Asian currencies threatened, on the one hand, to bring about a flood of goods into the Indian market from these countries and, on the other, to make it difficult for Indian manufacturers to compete in Asian and Western markets. Even so, India did not appear to face anything like the economic crises that engulfed South-east Asia and South Korea.

The Potential For Growth

Reforms easing the socialist and bureaucratic obstacles to foreign investment in India have made its business environment relatively investor friendly. With most regulatory restrictions removed, state governments are now competing with each other for foreign investment by offering the easiest and fastest access to their economies. The Indian government also points out that, in a survey of 274 Japanese companies by the Export-Import Bank of Japan, India was ranked third after China and Vietnam as among the most promising countries in the long term (ten years) for foreign investment. According to India's Ministry of External Affairs, an assessment of country risk ratings undertaken in 1995 by the UK's Economist Intelligence Unit, India was rated a lower investment risk than China, Turkey, Philippines, Mexico and Russia.

However, India's infrastructure (including banking and finance, water supply, power, telecommunications and transport) remains weak relative to the demands of its rapidly growing economy following economic liberalisation. In addition, the expansion of the private sector and its fast pace of growth have not been accompanied by a similarly rapid pace in the privatisation of several gigantic public-sector corporations. An estimated

$300bn in infrastructure investment may be needed over the next five years if GDP is to continue to grow at even its current rate, let alone the projected 8% by the turn of this century and 10% after that. A major effort is currently underway to develop the financial and non-financial infrastructure.

Potential, But Not Much Else

Although the newly elected government under the leadership of the BJP can be expected to provide only weak governance and may not survive for its mandated five-year period, the basic democratic and secular character of the political system is likely to persist. The radical shift from socialism to private capitalism and a market economy also now appears irreversible, although there may be some restrictions and a slowing down of foreign investment to protect domestic companies and small businesses.

Secessionist movements will continue to surface along the fringes, but are likely to dissipate eventually. The Kashmir issue, a major cause of contention between India and Pakistan, remains intractable, but the violent insurgency in the area has subsided since elections there in 1997. Despite reckless talk of another war between India and Pakistan in the early 1990s, this prospect has also faded. The fact that both India and Pakistan are covert nuclear weapon states may be one reason why the threat of a conventional war between them has faded. Given the prospect of peace in the region for the foreseeable future, India should be in a position to begin to realise its great potential. This, however, would require a strong government with a modernising outlook. There is no sign that such a government will be elected in India for many years to come.

◆

Africa

In 1997, Sub-Saharan Africa appeared to live up to its image of being a continent racked by violence and disorder. War either broke out or continued in what was Zaire, and is now the Democratic Republic of Congo (DROC), Congo-Brazzaville, Sierra Leone, Somalia, Senegal, Sudan and Burundi, while unrest erupted in Angola, Rwanda, Kenya, Central African Republic and Zambia. Although most African conflicts appeared to be driven largely by internal factors, a number of international trends are emerging. First is the growing influence of a Rwanda–Uganda–DROC axis that is also closely allied to Eritrea and Ethiopia. It has begun to play a dominant role in the affairs of Central and Eastern Africa. In West Africa, Nigerian troops have played an increasingly forceful role, heading the action in Liberia and restoring the democratically elected government in Sierra Leone (as part of the Economic Community of West African States Cease-Fire Monitoring Group)(ECOMOG). South Africa, despite facing domestic difficulties and rising crime rates, continues to be the economic and political driving force in southern Africa, and has extended its influence northward.

Although instability has been a characteristic of the region over the course of the year, a number of countries have shown solid progress and good rates of economic growth. Senegal's economy was expected to grow by 4–5%, and privatisation programmes in Côte d'Ivoire and Burkina Faso have begun to have a positive impact. Two countries that have recently emerged from conflict, Eritrea and Mozambique, have made particularly good progress. Eritrea's economy is forecast to expand by 10% in 1988. Mozambique has begun a wide-ranging programme to reform its banking sector, and has started to attract significantly more foreign investment and trade. Although it still relies heavily on overseas aid to make up around 60% of its income, most estimates indicate that this proportion is declining. Mozambique could be an early beneficiary of the World Bank's Heavily Indebted Poor Countries initiative from which it could benefit from a reduction in debt servicing of up to 80%.

South Africa's economy has been improving, and has attracted solid international investment. Foreign reserves have jumped to over $5.5 billion, and inflation, which was around 9% during 1997, is forecast to fall to 7% during 1998, the lowest it has been for a number of years. Nevertheless, South Africa's economy remains fragile. This was reflected in the 7% fall in value of the rand in August 1997, fuelled both by speculation that 79-year-

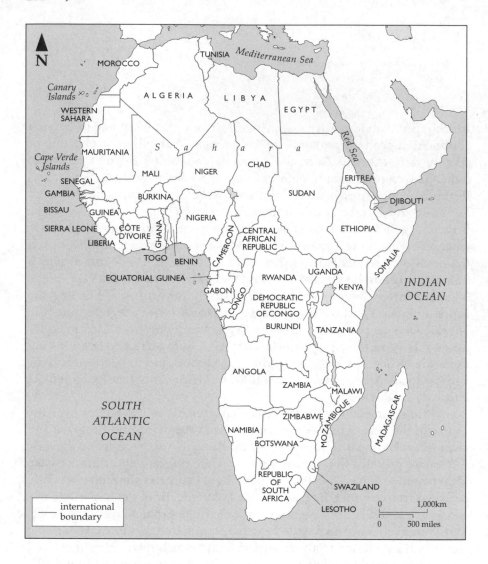

old President Nelson Mandela's health was failing and by changes to exchange control mechanisms. Although the government has generally followed orthodox economic policies aimed at redistributing wealth through high growth, the disparity between rich and poor – roughly translated into white and black – remains a source of friction among poorer sections of the population. Moreover, growth rates are unlikely to exceed 3% in 1998, not enough to address the large numbers of unemployed that have undoubtedly contributed to the serious rise in crime.

Inequity was one topic addressed in President Mandela's speech to the African National Congress (ANC) in December, when he announced his retirement as its leader. He criticised rich whites for clinging to their wealth and suggested that South Africa needed a 'system of social accountability for capital'. He also maintained the need for an 'African renaissance', something his successor Thabo Mbeki has also advocated. Mbeki, as the new ANC leader, looks likely to be appointed president of South Africa when Mandela steps down in 1999. The ANC was clearly relieved at the decision of Mandela's former wife, Winnie Madikizela-Mandela not to contest its deputy leadership position. Had she done so, there were dangers that the ANC would be split between her supporters and those who felt that she was unworthy of the position after substantial allegations were made against her at the Truth and Reconciliation Commission in November 1997.

Despite these positive signs, trends in Africa generally look bleak. The June 1997 UN Development Report notes that nearly 40% of Africans are poor and that this figure is growing; it is expected to reach 50% by 2000. Less than 5% of the world's private capital is invested in Africa, and the majority of what is invested is concentrated in the mining and extractive industries. Aid to Africa has also dropped significantly. The World Bank, for example, lent $1.74bn to African countries in 1996–97, down 37% from previous years. Crucially, the proportions spent on healthcare and education also declined sharply. World Bank policy has tended to favour those states that have followed economic reform programmes and more market driven policies. Surprisingly, some African nations support this concept. Ugandan President Yoweri Museveni criticised donor countries in March 1998 for not being more discriminating with their assistance, claiming that continuing to provide aid to 'dead economies' merely encouraged dependency.

Perhaps the greatest threat to progress is instability. Three strategically important countries – Kenya, Zambia and Zimbabwe – have been deeply affected by unrest. Kenya's increasing instability is unlikely to be improved by the chaotic elections held on 29–30 December 1997. The polls returned President Daniel arap Moi and his Kenya African National Union (KANU) party to power for another term. Organisational problems on election day were so bad that polling was allowed to continue for an extra day. Opposition groups claimed that election fraud had occurred and refused to recognise the result. Electoral irregularities certainly took place, including the siphoning off of state funds for KANU electioneering purposes. But Western ambassadors in Nairobi, whose governments had largely funded the election, recognised Moi's victory.

In reality, the main reason for his victory was the opposition's divisions rather than Moi's own popularity. As it was, over half of his cabinet lost their seats. Support for various candidates was largely split along tribal lines; many smaller tribes clearly preferring a *Kalenjen* leader to the prospect of rule by the largest tribal group, the *Kikuyu*. Despite his victory, Moi faces real economic difficulties. Assistance from the West, and from the International Monetary Fund (IMF) in particular, has been stalled until corruption is tackled. Given the economic difficulties and rising ethnic tension, Kenya's short term future looks set to be increasingly violent.

Zimbabwe, long regarded as a bastion of stability in the east, also faced periods of unrest. Food riots in January 1998 were the culmination of increasing frustration with President Robert Mugabe's leadership. Popular dissatisfaction was largely spontaneous, a response to the 36% rise in basic food prices. Although the reigning Zimbabwe African National Union–Patriotic Front (ZANU–PF) blamed trade-union agitators for the rioting, the problems can ultimately be traced back to poor economic leadership and political expediency. In an overtly political move to gain favour, Mugabe awarded pensions to war veterans, without allowing for them in the budget. He also announced a hugely popular land redistribution policy. The government would have taken over 4.7 million hectares of large commercial farms, mostly owned by white farmers, and given them to black farmers. However, the programme proved to be much too expensive and had to be abandoned. Other economic indicators offer little hope for improvement. Growth in Gross Domestic Product (GDP), forecast to be 6% in 1998, is not expected to rise above 2.5%; inflation is projected to run at close to 25% for the first part of 1998; and the Zimbabwean dollar, which was worth $0.41 in 1990, has fallen to $0.05. The IMF and the European Union have called for fiscal discipline. However, given that the present administration has held power since independence in 1980 it is doubtful that much progress will be made.

Zambia is one of Africa's more stable democracies. Nevertheless, a half-hearted military coup took place in October 1997 when two army captains captured the radio station and tried to incite a mutiny in the armed forces. The coup failed and President Frederick Chiluba has used the incident to conduct a wave of repression, imprisoning 84 middle ranking army officers and a number of opposition leaders. Also jailed without charge was Chiluba's old rival, former President Kenneth Kaunda. Despite heavy international pressure for his release and for human rights to be respected, Chiluba has remained intransigent. This hard-line stance has resulted in donor nations withdrawing loans, thereby undermining Zambia's potential budget surplus.

But for most of 1997–98, perceptions of Africa were shaped by geopolitical changes in Central Africa where new alignments between countries were taking place and their leaders were developing new strategic interests. The DROC is central to future developments in the region. Its size and common border with nine other African states means that Central Africa's future is linked to its continued stability and, to a large degree, its economic prosperity.

Profound Changes In Central Africa

There has been a fundamental change in the political and military situation in Central Africa. The most significant strategic shift in power in the region was the collapse of Mobutu Sese Seku's 32-year reign as President of Zaire, now known as the Democratic Republic of Congo (DROC) and the inauguration as president of Laurent Kabila, who headed the Alliance of the Democratic Forces for the Liberation of Congo-Zaire (ADFL). The ADFL advance faced little opposition from Mobutu's corrupt and underpaid army. With its collapse, the capital Kinshasa was taken in May 1997, just six months after the start of Kabila's military campaign.

The change of power was widely touted as a new beginning for a country that had been crippled by Mobutu's corrupt rule. But this early optimism has been tempered by the recognition of the enormous obstacles which, from the very beginning, faced development in the DROC. Immediately following the ADFL's military success, the new regime was accused of massive human rights abuses. It was blamed for the disappearance and death of thousands of Rwandan Hutu refugees who had chosen to flee west instead of joining the exodus east back to Rwanda when the ADFL advance closed down the refugee camps in eastern DROC. The incidents were widely believed to have been orchestrated by Rwanda, which wanted to avenge the loss of life from the 1994 genocide and was reluctant to import Hutu hardliners back to await trial in Rwanda's overcrowded prisons. A UN human rights team sent to investigate the killings was prevented from travelling to the region for several months because of the DROC government's disagreement over the period and scope of the inquiry. The long delays were inevitably believed to be stalling tactics to enable Kabila's government to hide the evidence of any atrocities.

The US, which supported Kabila's rise to power, found itself with a difficult policy dilemma. US backing had been primarily aimed at removing the corrupt and malingering Mobutu who, while a staunch Cold War ally, had resisted any democratic change. Some in the US Congress and government wanted to follow the lead of African states (notably, Rwanda, Uganda and, to some degree, South Africa), and give Kabila time to put his house in order. Taking a hardline stance, it was felt, might limit future leverage and push him further towards authoritarian rule. On the other hand, dismissing human rights abuses on such a scale feeds the 'cycle of impunity' that has been a feature of human rights in the region.

If the US threatens to withdraw development assistance until the DROC government's human rights record improves, it runs a serious risk of losing influence with the Kinshasa administration. DROC is not without other options. For the first time, the private sector has been actively courted to play a central role in the country's future development. The US

Bechtel Corporation, for example, has offered to design a development strategy free of charge, possibly eclipsing World Bank and IMF initiatives. Of interest to Bechtel is the DROC's immense mineral wealth of cobalt, copper, gold and diamonds. The company would be able to claim an unprecedented access to mineralogical surveys, both charted and yet to be mapped, and be strongly placed for the allocation of mining rights that have been actively contested since Mobutu's departure. Alongside these commercial interests, the EU's more hardline stance against human rights excesses and threats to postpone the delivery of humanitarian assistance did not appear to carry much weight.

Kabila also faces serious domestic opposition. The ADFL could easily fragment. Vice-President Masau Nindingu was arrested in November 1997, allegedly for forming his own militia. The Mai Mai ethnic group, formerly part of the ADFL, has mounted attacks against government positions and there are reports that it has links with the remnants of *interahamwe* guerrillas that carried out the 1994 Rwandan genocide. Etienne Tchisekedi wa Malumba, one of Kabila's main opponents, was hounded out of Kinshasa and has been whipping up support from his home areas of Kasai. This underlines another problem. Under Mobutu, several of the provinces had been effectively autonomous, because the central administration had collapsed. Kabila now threatens this autonomy; the relationship between the central and provincial administrations will need to be carefully managed. The new regime also suffers from the perception that it is dominated by easterners from the Kivu areas and maintains its close links to Rwanda and Uganda.

Angola's Sphere Of Influence

The change in power in Kinshasa affected other central and southern parts of the continent. The DROC's neighbour, Congo-Brazzaville, erupted into conflict just one month after Kinshasa fell. The war, sparked by President Pascal Lissouba's attack in June on his rival, former President Denis Sassou Nguesso, appeared to be entering a stalemate while a death-toll of over 10,000 continued to mount. However, the tide turned in Nguesso's favour on 15 October 1997 when Angolan forces, under Chief of Staff, João de Matos, crossed into Congo-Brazzaville from the Angolan enclave of Cabinda. The conflict provided Angola with an opportunity to close off another source of support for Jonas Savimbi's rebel group *União Nacional para a Independência Total de Angola* (UNITA) and to extend its sphere of influence northwards. Angola had previously supplied Kabila's forces with weapons and other military support – there were frequent reports of Angolan soldiers fighting alongside the ADFL against Mobutu's forces – for the same reason. Under Mobutu, UNITA had enjoyed unrestricted access across the northern Angolan border into Congo to sell diamonds – believed to be worth approximately $500 million a year – and to buy weapons and supplies. With the change in government in both the DROC and Congo-Brazzaville, Angolan President José Eduardo dos Santos has achieved considerable strategic advantage over his old enemy. The stability that has thus been obtained has also guaranteed the protection of his own oil revenues.

Although Angola's incursion over the border drew criticism from the UN, there appears to be unofficial support for an action that resulted in Congo-Brazzaville's first democratically elected president being ousted. For example, Nguesso was quickly recognised by France, which was anxious to ensure that French oil company Elf-Aquitaine's interests in the rich oil reserves of Pointe Noire remained secure. But Nguesso's plan for a long transition period before holding new elections has embarrassed Western governments, which acknowledge that, apparently at least, the events have set a bad precedent for democratic change in Africa. In the meantime, Nguesso's lack of any real popular support means that Angolan forces will need to remain in Congo-Brazzaville for some time to provide security for his regime.

Angola's hand has also been strengthened by growing international disillusionment with UNITA's stalling on the timetable for demobilisation. It was supposed to have fully demobilised its soldiers by February 1998. The deadline was not met. Savimbi's attempt to blame the inefficiency of the UN did not impress the three outside negotiating powers, Portugal, Russia and the US. However, the Angolan government took it upon itself in March to legalise UNITA, giving it full party-political status as part of moves to advance the stalled peace process. The question is whether UNITA will meet the new timetable to demobilise. It still has sufficient forces to remain a guerrilla threat in Angola's dense bush, but it has isolated itself from its former allies and removed its chances of becoming a mainstream political actor. The Angolan government has also warned Zambia against allowing UNITA to use its territory to ship arms and supplies. In what amounted to an open threat, Luanda noted the recent changes in power in Congo and Congo-Brazzaville, and warned Zambian President Frederick Chiluba that similar events could happen in his country.

Rwandan Gains?

Much of the military success of the ADFL in 1997–98 has been attributed to Rwanda's acknowledged role in supplying troops and providing tactical leadership. Rwanda's support for Kabila appears to have had two objectives. The first was to close the refugee camps on its borders, force the return home of the refugees and, with their return, gain access to the organisers of the genocide in 1994 which killed around 800,000 Tutsis and moderate Hutus.

The second, less obvious, goal appears to have been a desire to create a zone of influence in north and south Kivu, the DROC provinces that border Rwanda. Much of the population in this area is of Rwandan origin, including a number of ethnic Tutsis, locally known as *Bangamulenge,* who spearheaded Kabila's advance. Diminishing land resources in Rwanda have long been cited as an underlying factor in the conflict. Rwanda is the most densely populated country in Africa, and individual landholding has declined rapidly as populations have increased, reducing the country's ability to meet the subsistence needs of the population. The change in power across the border may now give Rwanda the space it needs for its growing population. International frontiers are likely to remain unchanged, but contiguous sympathetic governments might review their nature, allowing freer movement of people and trade than was possible before. Ugandan President Yoweri Museveni has long argued for a regional economic union that might overcome the artificiality of many Central African borders.

Yet Rwanda's situation in the medium term appears untenable. Rather than resolving what Rwanda believed was its greatest threat, the influx of more than one million refugees from the camps in Zaire and Tanzania at the end of 1997, has imported the violence within its borders. Attacks on government officials and Tutsis have increased dramatically and are no longer confined to the western areas of the country bordering the DROC. These attacks have been met with aggressive cordon and search operations by the Rwandan military that have largely targeted the Hutu population. The operations are believed to have resulted in hundreds of deaths each week, prompting many Hutus to return as refugees to Tanzania. In addition, more than 120,000 people await trials in the country's grossly overcrowded jails. Despite its best efforts, the justice system has been unable to cope with the numbers. At current rates of processing, many of those held in custody could be waiting over 25 years for a trial date.

For all the seriousness of Rwanda's situation, it has not reached the depths plumbed by Burundi. Since seizing power from Hutu President Sylvestre Ntibantunganya in July 1996, Burundi's leader, Major Pierre Buyoya, has made few gains. Burundi supported Kabila's victory, thereby removing from its western border the opposition Hutu military bases from which were mounted attacks into Burundi. But fighting has continued. Most of the opposition has been under Interior Minister Leonard Nyangoma, who heads the *Forces pour la Défense de la Démocratie* (FDD) which operates both within Burundi and unofficially from bases in Tanzania.

Buyoya claims that he wants to unite all non-violent parties. But his own policies are extreme. Throughout 1997, several hundred thousand Hutus were resettled under a 'villagisation' programme into camps under Tutsi army protection. The intention was to prevent the movement of Hutu guerrillas amongst the population. People are permitted to leave the camps periodically to tend fields, but health and general living conditions are poor and there is now the threat of famine as a result of reduced food production. Inevitably, Hutu attitudes have hardened towards the Tutsi-led regime.

Buyoya also faces opposition from within his own ranks. Extremist Tutsi 'supporters', many of them businessmen who have suffered from the war, urge an even more aggressive response. Buyoya's imprisonment of arch rival Baptiste Bagaza and other army officers shortly after taking power have forestalled attempts to overthrow him. The split between the two men is so extreme that Bagaza has begun limited negotiations with Hutu groups in an attempt to isolate Buyoya, and he is reported to have argued for a settlement based on the separation of Hutu and Tutsi geographic areas. Countries in the region have pushed for a wider dialogue with opposition groups and, in September 1997, the East African Heads of State meeting approved the extension of regional sanctions. Burundi has managed, despite these sanctions, to continue exporting coffee, its main source of foreign exchange. If there has been a decline in the production of coffee and tea that has adversely affected the economy, it was a result of the government's own repressive policies.

The New Leaders

Kabila's ascension has wider significance in that he is a representative of a handful of leaders that represent the new African nationalism. At the centre of this grouping is Ugandan President Yoweri Museveni, who came to power in a guerrilla campaign against Milton Obote in 1986 and who provided support to Paul Kagame's Rwandan Patriotic Front (RPF) which ousted the Hutu-led government in Rwanda in 1994. Wider backing for Kabila was also supplied by Eritrean President Issaias Afewerki, who led the successful struggle for independence from Ethiopia, and Meles Zenawi, Prime Minister of Ethiopia, whose forces ousted dictator Mengistu Haile Mariam. Experienced military campaigners now turned politicians, these individuals are shaping a new style of post-colonial Africa which is in marked contrast to the approach of earlier heads of government who leaned heavily on the West. The new group is fiercely independent in outlook and has resisted pressure to adopt Western models of democracy. (In Uganda, Museveni has rejected a multiparty system because, in his view, it favours tribalism.) These leaders have followed pragmatic economic policies that include a degree of pri-vatisation and, in what is becoming another hallmark, have demanded greater accountability from international non-governmental organisations (NGOs), thus strictly limiting their activities. A number of NGOs have even been forced to leave Rwanda and Eritrea.

To a large degree, this grouping – Eritrea, Ethiopia and Uganda particularly – are also bound by common strategic interests, notably their shared border with Sudan and opposition to its ruling National Islamic Front (NIF). Zenawi adopted a more aggressive policy towards Sudan after the Sudan-backed attempt to assassinate Egyptian President Hosni Mubarak in Addis Ababa in 1995 and has provided direct support for John Garang's rebel Sudan People's Liberation Army (SPLA). In addition to facing opposition on its borders, the NIF is also under threat from within. For the first time in the history of the civil war, northern and southern forces have joined together to oppose the Khartoum administration. The US, which accuses Sudan of supporting international terrorism, announced sanctions against it in November 1997, having the previous year given $20m in military assistance to Eritrea, Ethiopia and Uganda for 'defensive purposes'. Most of this aid was believed to be used by Sudanese opposition groups, particularly the SPLA. The war in general has been going against the NIF, with the SPLA advancing steadily on the southern towns of Juba, Kassala and Kadugli. If these towns were to fall, the course of the war would change dramatically, giving impetus to the inconclusive peace talks dragging on in Cairo.

The US has been quick to recognise and encourage the new breed of African leaders, and their relations with the US have tended to be very favourable, with a consequent enlargement of US influence in the region. US Secretary of State Madeleine Albright's six-nation visit to Africa in December 1997 emphasised this. She met with heads of state from the DROC, Ethiopia, Rwanda and Uganda, and concentrated in her talks on shared regional security objectives while carefully avoiding any stress on human rights and democracy issues. President Clinton followed this up with a ten-day, six-nation tour (Ghana, Botswana, Rwanda, Senegal, South Africa and Uganda) that began on 24 March. It was the first visit to Africa by a US President in 20 years.

France characterises Washington's policies as a 'conspiracy' to limit French influence in the region. France has found it more difficult to forge good relations with the new leadership because of traditional ties with the old regimes. Its support for the regime in Rwanda in 1994, and its subsequent advocacy of an international military intervention into the DROC in 1996 that would have slowed Kabila's advance, has put it at odds with the new administrations. France has made some efforts to improve relations with the new regimes, and Kinshasa has also tempered its opposition as a means to win EU funding. But there is little doubt that US commercial interests have benefited from the change in power in the DROC. American Mineral Fields, for example, has acquired a $1bn concession to exploit copper and cobalt in the Lubumbashi region in the south-east of the country.

Nigeria Flexes Its Military Muscle

Nigeria has actively engaged in the affairs of its conflict-plagued neighbours, even demonstrating a willingness to intervene militarily. It has undoubtedly strengthened its position in the region, which has been interpreted by some observers as evidence of a growing Nigerian hegemony. Its action in February 1998, when it attacked the military junta in Sierra Leone and pushed Major Johnnie Paul Koroma's Armed Forces Revolutionary Council (AFRC) military forces out of the capital, Freetown, was widely praised. The AFRC had seized power in a military coup in May 1997, ousting a civilian government headed by Ahmed Tejan Kabbah which had been elected just a year before. The Sierra Leone Army quickly consolidated its grip on power by joining forces in an unholy alliance with the Revolutionary United Front (RUF) which it had been fighting for the previous seven years. This completely reversed the fortunes of the RUF, which had suffered military defeats at the hands of the security forces backed by South African military company, Executive Outcomes (EO), that the Sierra Leonean government had hired to fight its battles. Koroma argued that his seizure of power was to combat corruption and increasing tribalism; Kabbah came from the Mende tribe from the south, whereas most of the army were from the north and east of the country. A more plausible explanation is that the army was angered at Kabbah's plans to reduce the power of the military, including the removal of privileges such as extra rice rations, and to exclude it from diamond-mining areas that, in the past, had provided a lucrative addition to meagre salaries.

Although Koroma's coup attracted widespread international condemnation, Nigeria was the only nation ready to back sanctions with military force. An earlier clash between the AFRC regime and a small Nigerian contingent, based in Sierra Leone to guard key installations, increased the chances of a military outcome. However, the prospect of Nigeria taking a forceful lead to back a democratic government given its own woeful record on democracy was not only ironic, but also presented Western nations with a difficult dilemma. Led primarily by the US and UK governments, they had sought to isolate Nigeria's military dictator, General Sani Abacha, and had pushed for democratic reform. The Commonwealth Heads of Government Conference in Auckland in 1995 had suspended Nigeria from the Commonwealth. The Edinburgh Commonwealth meeting held in October 1997 considered that little had changed as far as democracy and human rights were concerned and voted to maintain the suspension, clearly angering Nigeria. Re-admittance will depend on whether the Nigerian elections, due to be held in October 1998, take place and are free and fair.

Nigeria's military interventions have also strengthened its standing in the 16-member Economic Community of West African States (ECOWAS). In

March 1998, ECOWAS officially praised Nigeria's role in the return of the Kabbah to office in Sierra Leone in February. It also decided to make the West African Ceasefire Monitoring Group (ECOMOG), the military arm of ECOWAS, a standing arrangement. Previously it had been constituted only in response to given situations. As Nigeria has provided the bulk of ECOMOG's troops, this too will strengthen its position in the region. Unease remains, however, amongst francophone countries which have traditionally attempted to limit Nigeria's influence. As some states face their own internal problems, they have pushed for strict guidelines governing when an ECOMOG force might be deployed, and its exact legal status.

Nigeria's performance in Sierra Leone followed closely its role in the resolution of the seven-year Liberian conflict. Elections held in July 1997 were won overwhelmingly by Charles Taylor, head of the National Patriotic Front of Liberia (NPFL), who took nearly 75% of the vote. The elections were made possible largely because Taylor and Nigeria had finally reached an agreement. Yet Nigeria has hardly been impartial in the conflict and must share some of the blame for its protracted nature. Conflict between the two had frustrated earlier efforts to reach a settlement. Taylor was bitter that an ECOMOG intervention into Monrovia in 1990 had prevented him from winning the country militarily. During the stalemate that ensued, Nigeria had played both the role of mediator and antagonist, and was accused of arming rival Liberian factions against the NPFL. It had also become heavily involved in commercial activities, particularly the export of hardwoods, despite an economic embargo on Liberia.

There are some optimistic indications that the election outcome might end a civil war that has claimed the lives of over 150,000 people. Despite the positive political result, however, Liberia's economy is in ruins. Government structures and institutions ceased to exist shortly after fighting began. In such dire circumstances, the disarmament and rehabilitation of 25,000 'fighters' from the various factions is problematic. The Krahn tribe, which once formed the core of the armed forces and constituted a significant proportion of three other factions, are also nervous about their survival with the NPFL in power. ECOMOG, led by Nigeria, has remained in Liberia, ostensibly to oversee the demobilisation process. The accord between Taylor and the Nigerians may be breaking down. Taylor, for example, had backed the RUF in Sierra Leone against the wishes of ECOMOG.

Western Disengagement Continues

Although Western powers are concerned that Nigeria's image has been given a false lustre by its action in Sierra Leone, the emergence of a regional African power capable of performing peacekeeping and enforcement operations is generally welcomed. There is little appetite in the West for intervention in African affairs. Even France, which has remained the most interventionist-minded, is reducing its support to former colonies. It announced in December 1997 its decision to reduce both troop numbers and the number of French military bases (see map, p. 242). First to be targeted is the Central African Republic, where French soldiers intervened during an army mutiny in January 1997 to save President Ange-Félix Patassé's regime. The total number of French troops is scheduled to decline by around 40% to some 5,000 by the year 2000, reflecting waning economic and strategic interest in the continent.

In its place is a policy of 'African solutions to African problems' – supporting African states in resolving their own conflicts. In reality, this is more a policy to ensure Western forces are not called on to provide military intervention in the future. When Warren Christopher was Secretary of State, the US initiated the African Crisis Response Initiative, whereby the US would train and equip up to ten battalions of troops from a similar number of African states that would be ready to act jointly in a humanitarian emergency. Most training has concentrated on adopting a set of common operating principles and equipment. The force remains in its infancy, and will still depend on Western financial support, both for its training and, ultimately, for deployment. A budget of $35m, of which the US will contribute $20m, will first train troops from Senegal, Uganda and Malawi. Similar plans will then target Mali, Tunisia and Ghana. But the initiative continues to be hindered by difficulties in its earlier stages. The UK and France had launched their own plans to train African armies in peacekeeping techniques, and it was unclear to some how the US initiative would fit with these other approaches. Meanwhile, some African nations were concerned at what they believed to be US unilateral action and insisted that the peacekeeping force should act only with a UN mandate and be incorporated into UN peacekeeping structures.

Africa will largely be left to deal with instability itself in the future. Yet Nigeria has shown that it is prepared to act unilaterally, even if it flouts international opinion. South Africa has been a more reluctant actor, despite US urging. It has largely focused on its own domestic situation, including the reform of its military with the merging of the forces of the African National Congress (ANC) and the old South African Defence Force. The African Crisis Response Initiative, despite US optimism, is likely to be constrained by a lack of resources and the restricted situations to which it can be deployed. Instead, there are signs that a tougher line is being taken by the World Bank and IMF against corrupt regimes. The IMF has refused Kenya a $200m loan, for example, until it tackles its endemic corruption. Using economic leverage can bring administrations into line and produce positive results in the longer term. The risk, however, is that if fragile regimes are left more unstable, the same policies may have violent consequences.

Anguish In Algeria

Algeria was often in the headlines in 1997 because of the frightful violence that affected the country throughout the year. The security situation was so complex and the sources of violence so many, that outsiders were at a loss to know how to influence the situation in a positive way. Calls for improved dialogue between the government and opposition groups were about the limit of what Western powers were willing or able to do to shape events. This meant that there was little outside influence on the country's military government as it concentrated on its three-year project to recover political legitimacy and contain the electoral opportunities of Islamic-dominated groups.

Ostensibly, President Lamine Zeroual's objective has been to ensure a strong context in which limited political pluralism – even Western-style democracy – can exist, based on an extraordinarily strong presidential institution and a directly-elected parliament, with a carefully constituted upper house providing a further balance.

In reality, the project's aims are to prevent a repeat of the political crisis that began with the clear win by the Islamic Salvation Front (FIS) in the first round of parliamentary elections in December 1991 leading to the military 'coup' of January 1992. Widespread charges that the unwritten agenda of Zeroual's project has been to preserve the unofficial dominance of the military-political élite, and prevent the Islamist opposition from ever repeating its victory of 1991, are hard to counter. Participation in debate and government has only been extended to the regime's traditional supporters, and the mass of the population, particularly the young, has been marginalised.

Playing The Democracy Game

The project's early successes, most notably Zeroual's clear win in the November 1995 presidential polls, have been followed by a steady increase in public frustration at the President's apparent inability to deliver his central pledge to ensure security in return for strong government. At the referendum to approve changes to the constitution in November 1996, the opposition claimed that turnout in some traditionally Islamist areas was as low as 15%. This contrasts sharply with the official national vote in favour which was more than 85%.

One of the shrewdest elements of the programme has been to lay down strict conditions under which political parties can register, and therefore contest elections. While on the one hand this is designed to inhibit regionally-based parties (regionalism is one of the factors on which the internal dynamic of the regime depends), on the other it works in favour of the largest fronts that enjoy national or near-national organisation and whose platforms have traditionally broadly coincided with the interests of the ruling élite. This has included, since its long-awaited foundation in March 1997, the National Democratic Rally (RND), as well as the former socialist National Liberation Front (FLN) and the 'conservative' (nowadays a code for 'cooperative Islamist') Movement for a Peaceful Society (MSP).

Under the amended constitution, all political parties must hold founding congresses attended by 400–500 delegates elected by more than 2,000 supporters from at least 25 of Algeria's 48 provinces. The constitution also bans any party whose platform is explicitly based on religion, thus justifying the continued prohibitions on the FIS and similar organisations, including the Movement for Democracy in Algeria (MDA) headed by the country's first president Ahmed Ben Bella. However, application of the regulations has not been entirely consistent, and their greatest value to the government lies in their potential to be used against political organisations that pose too great a challenge to the status quo.

The RND was more or less openly conceived as a modernised version of the FLN, a political vehicle in which the 'reformist' technocrats and their supporters would feel most comfortable. Its platform is similar to those of many similar political fronts in developing countries, but its primary function is to legitimise the political authority of the third and fourth generation of Algerian politicians and administrators since independence, whose interests appear – at least for the time being – to be best advanced by Zeroual's project. This was soon underlined, as almost all of the pre-election cabinet of the Prime Minister, Ahmed Ouyahia, joined the new party. It rapidly established a nation-wide organisation, greatly assisted by its association with the two most powerful non-political organisations, the General Union of Algerian Workers (UGTA) and the war of independence veterans' association.

Algeria's first ever completed 'democratic' parliamentary elections took place on 5 June 1997. It is impossible to assess accurately the extent of irregularities in the polling, in which the RND officially won 33% of the vote, the MSP 15% and the FLN 14%. Seven other parties and independent candidates won the remaining seats. The official turnout figure of 69% was almost certainly exaggerated and designed to add credibility to the election process. As expected, Ouyahia was reappointed prime minister and formed a three-way coalition of 'national consensus' between his RND, the FLN and the MSP. Although Algeria undeniably enjoys a relatively advanced level of genuine parliamentary and, to some extent, public debate over some central issues, the coalition does not represent any real change from the previous unelected government. Zeroual's main objective has been to strengthen the impression of a national consensus behind the regime's hardline security stance. The MSP controls several portfolios that can be considered poisoned chalices, such as housing, and which are perhaps intended to draw popular frustration away from the RND.

A Murderous Atmosphere

Algeria has made some undeniable strides towards the establishment of a functioning political pluralism, but the influence of the military establishment remains as powerful as ever. As in countries such as Pakistan and Turkey, the political apparatus operates under the unarticulated threat of a military veto, and at worst a military intervention. Although ostensibly elected democratically in 1995, President Zeroual's main strength is his own military background and links with the senior officer class. The relative strengths and opinions of these officers, however, are notoriously difficult to interpret, although during the year the more moderate elements who tend to cluster around Zeroual appeared to win some key victories. Notably, in October the President appointed a supporter, General Boughaba Rabah as commander of the First Military Region (Algiers), the most important in the country.

Ironically, an attempt on the part of the new government to improve international awareness of Algeria's internal problems precipitated an unprecedented scrutiny of the regime's human rights record and sowed the seeds of moves to 'intervene' in the country. The government allowed an unusual number of foreign journalists to cover the local elections of October 1997, in which, not surprisingly, the RND comfortably won most seats. The media not only roundly accused the government of large-scale fraud (with some justification), but also focused on allegations that members of the military and its associated agencies had played a direct role in a series of staggeringly brutal raids on civilian villages.

These raids, in which occasionally more than 200 villagers are slaughtered in up to five hours of systematic butchery, have generated the term 'Algerian-style terrorism'. The worst affected areas are the Mitidja plain, the densely-populated agricultural area to the south of Algiers, and more remote areas such as the mountains of the north-west (see map, p. 252). In one of the worst instances, up to 300 villagers were murdered in the agricultural community of Sidi Rais, just 12 miles south of Algiers, on 28 August 1997. As is the pattern in many attacks, the attackers wore stolen army uniforms, and there was no response from a nearby police post during the attack.

Various motives have been suggested: that the most active extremist entity, the Armed Islamic Groups (GIAs), are engaging in 'straight-forward' terrorism; that the GIAs are wreaking revenge on formerly supportive villagers who have redirected their support to the state; that elements within the state are engineering the attacks to ensure public support for the regime's hardline security policies; that former landowners are employing guerrillas to drive villagers from their lands; or that government-supplied self-defence militias are engaged in vendetta violence. The truth probably lies in a combination of all of these, but is further complicated by a high level of criminality that flourishes in a highly opaque situation.

The raids were merely the most noticeable element in a tide of violence that has claimed at least 60,000 lives since January 1992, though tight state control and a terrorised population mean that any attempt to quantify the extent of the violence is inevitably flawed. The GIAs carry out bombings in public places, assassinations and kidnaps intended to 'remind' the population of God's law, as well as the aforementioned raids on villages. Reports of almost unimaginable brutality against women, children and the old became commonplace during 1997, adding weight to the view that vendetta violence explains many attacks.

Between 1993 and 1996 the GIAs reportedly murdered almost 120 foreigners, most of them long-term residents of Algiers, although no such incident occurred in 1997. There have been no convincing explanations why Algeria's crucial oil and gas sectors have been largely unscathed by the violence across the north. There are populations that can be easily infiltrated and migrant work-forces with proven opposition sympathies involved in the sector, although most locations are remote and the security forces are reliable. There were at least three sabotage attacks against oil and gas pipelines during 1997, but no direct attack against foreign expert or semi-skilled workers that, if it took place, would surely jeopardise critical international confidence.

However, there was an important change in the balance between the state and the GIAs and their sister front, the FIS' armed wing, the Islamic Salvation Army (AIS), during 1997. State policy since 1992 has been gradually to push the terrorists and guerrillas out of easily controllable and strategically important towns and transport routes into the mountains, uplands, desert fringes and densely-populated cities that are almost impossible to secure. Inferior counter-terrorist techniques, combined with a high level of popular support for the extremists among remote populations, inhibited this until 1997, when the security forces began to regain the upper hand. By mid-year, the security forces had cleared all areas of the GIAs and the AIS except Greater Algiers and the adjoining Mitidja Plain area, the areas around Tlemcen and the north-west, and parts of the north-east.

The disparate elements using Islam as their banner do not co-ordinate in any way, and no single individual could deliver a peace agreement with the government even if one were on offer. There have been attempts, however: Zeroual ordered the release of Abbasi Madani and Abdelkader Hachani, respectively the leader and number three of the FIS, in July 1997, though both have been placed under house-arrest for infringing strict bans on 'political activity'.

It remains very unlikely that the regime will legalise the FIS again or reintegrate it into political life. The release of the leaders was a calculated gamble primarily designed for an international audience: both are less important than the FIS' number two, the firebrand preacher Ali Belhadj, who has served six years of a 12 year sentence. Their release did, however, lead to the AIS' decision to cease hostilities in October 1997, though this appeared to be in danger of being reversed on several occasions over the following months.

Unfortunately for the Zeroual regime, the completion of the political project, the releases of Madani and Hachani and the AIS cease-fire were almost entirely overshadowed by growing international outrage at Algiers' human rights record. As a suspicious succession of individuals claiming to be defecting security officials condemned the Algerian regime in Western newspapers for a wide range of abuses, Western governments began to urge Zeroual to let them assist or intervene. In response to the government's widely suspected fraud in the local elections of October 1997, the European Union sent the first ministerial delegation to 'investigate' the security situation and alleged human rights abuses in January 1998, though, as expected, without producing any significant result.

Adjusting Economic Life

Diversification away from over-dependence on oil and gas revenues remains a prime objective of the Algerian government. The centrist economic policies of Houari Boumedienne, president from 1965 until 1978, established a state industrial sector that was characterised by inefficiency, mismanagement and over-employment. From being a largely agricultural society before independence, Algeria is now forced to import most of its food. Zeroual and Ahmed Ouyahia have made structural adjustment and economic reform key priorities and, in the face of opposition from powerful entrenched interests, are making striking progress in some areas.

The privatisation of state enterprises is naturally one of the main pillars of these initiatives. The government has started a programme to sell off up to 230 state-owned companies, either totally or partially. Firms in the construction, major public works, building materials, tourism, distribution networks, chemicals and pharmaceuticals sectors are to be privatised. The consolidation of the first stage of this programme took place in 1997, as enterprises were absorbed into 11 holding companies that will ensure improved management, efficiency and sales, as well as significant reductions in the workforce.

Algeria desperately needs skills and technology to upgrade facilities that were mostly established in the 1960s, but the government has a difficult task in persuading international investors outside the hydrocarbons sector to look at the country. A central component of the government's strategy involves a foreign company entering into a partnership with a state-owned enterprise to upgrade existing production facilities. The local firm benefits by getting new technology and an upgrade to international standards, while the foreign company gains access to the market.

Concerns over political risks and the security situation clearly inhibit international investor interest, although this has steadily deepened since the August 1996 London Club rescheduling of $3.2 billion of the country's $4.5bn commercial debt. Robust support from the World Bank and the International Monetary Fund prepared the ground for the rescheduling, while reductions in the inflation rate and the initial stages of economic restructuring have removed the need for a further rescheduling, at least until 1999–2000.

However, many underlying economic figures remain unsettling. Algeria's total debt was estimated in late 1997 at around $33.5bn, up from $22.5bn in 1994. The unemployment rate hovers around 28%, and in some worst-hit urban areas it is closer to 60%. The government has made the first real efforts to tackle the chronic shortfall of around two million housing units, which is traditionally viewed as one of the prime factors behind the popularity of the radical opposition. Elected local authorities have been granted the main role in distributing state-subsidised housing to poor and lower middle-class citizens. Oil price fluctuations determine not only the overall health of the Algerian economy, but also the likelihood of a further debt rescheduling.

The oil and gas sector remains the star attraction, however. More than 40 foreign oil companies, drawn mainly from the US and Western Europe, have taken advantage of liberal production-sharing agreements with the state oil and gas corporation Sonatrach since 1991, and many more will sign up over the coming years. The government has opened up the downstream hydrocarbons sector to international investment, and begun to restructure Sonatrach itself, which has long enjoyed a reputation as the best-managed organisation in the country. This was the first Algerian entity to issue a corporate bond, in early 1998.

Algeria's place in the world presents another difficult challenge for the government. Relations with France remain critical, despite the first signs of a commercial reorientation towards the US because of the high involvement by US oil companies in Algeria. The coming on-line of the Euro-Maghreb gas pipeline in 1997 forces the governments of the southern EU to take a close interest in events in Algeria, and this will only increase as the gas requirements of Spain, Portugal, France and Italy increase over the next ten years.

Facing An Uneasy Future

Zeroual's political project has exceeded expectations in resolving the crisis that was immediately triggered by the FIS' electoral victory in 1991, but which goes back in many ways to political and social questions left unresolved at independence. However, it is essentially a short-term compromise. The framework will face its next real test in 2000, when the second presidential elections under Zeroual's rules will take place. Zeroual's prime challenger is likely to be Mahfoudh Nahnah, the veteran 'conservative' leader of the MSP, whose low profile during the 1997 parliamentary and local elections was intended to consolidate support for 2000. Few doubt that the military would be prepared to intervene if the result was not to their liking.

In addition to its security problems, the government faces the challenge of managing the transition to a more liberalised economy without triggering major social unrest. Organised trade unions are one of the backbones of the RND and the FLN, and, inevitably, large cuts in the workforce could undermine this critical relationship between workers and the government, and fuel strikes against Ouyahia's policies. Five years of union silence ended in late 1997 as the main General Union of Algerian Workers (UGTA) threw down the gauntlet to the government, and organised isolated strikes in February 1998. A sustained period of industrial unrest could be generated by a general strike such as the one called by the FIS in 1991 which brought the military to demand the resignation of the government. There is the risk that another would again draw the unwanted attention of the military and destroy any progress which the present leaders have made.

◆

Strategic Geography 1997/98

———	international boundaries	*Dar El Beida* ⊕	international airport
——·—	province or state boundaries	*Al Kharj* ⊕	air base
SONORA	province or state	*Tengiz* ▨	oilfield
————	disputed boundaries	⬕	oil refinery
▣	capital cities	⁀⌒	rivers
●	cities/towns	▱	lakes
◣	built-up areas	▲	mountain peaks

Foreign military presence in the Middle East

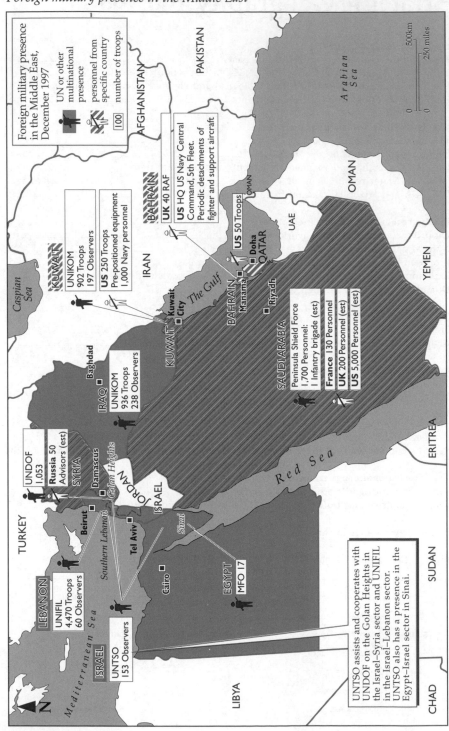

Foreign military presence in the Middle East, December 1997

UN or other multinational presence

personnel from specific country

number of troops

100

KUWAIT
UNIKOM
902 Troops
197 Observers
US 250 Troops
Pre-positioned equipment
1,000 Navy personnel

BAHRAIN
UK 40 RAF
US HQ US Navy Central Command, 5th Fleet. Periodic detachments of fighter and support aircraft

US 50 Troops
US Doha
QATAR

BAHRAIN
Manama

Riyadh

SAUDI ARABIA
Peninsula Shield Force
1,700 Personnel:
1 Infantry brigade (est)
France 130 Personnel
UK 200 Personnel (est)
US 5,000 Personnel (est)

IRAN

The Gulf

Kuwait City

KUWAIT

Baghdad

IRAQ
UNIKOM
936 Troops
238 Observers

SYRIA
UNDOF
1,053
Russia 50 Advisors (est)

Damascus

Golan Heights

JORDAN

ISRAEL

Beirut

Tel Aviv

Southern Lebanon

Sinai

LEBANON
UNIFIL
4,470 Troops
60 Observers

ISRAEL
UNTSO
153 Observers

Cairo

EGYPT
MFO 17

UNTSO assists and cooperates with UNDOF on the Golan Heights in the Israel–Syria sector and UNIFIL in the Israel–Lebanon sector. UNTSO also has a presence in the Egypt–Israel sector in Sinai.

AFGHANISTAN

PAKISTAN

Arabian Sea

OMAN

OMAN

UAE

YEMEN

ERITREA

Red Sea

SUDAN

CHAD

LIBYA

EGYPT

Mediterranean Sea

TURKEY

Caspian Sea

0   500km
0   250 miles

N

Build-up of Allied Forces during the UNSCOM inspection crisis

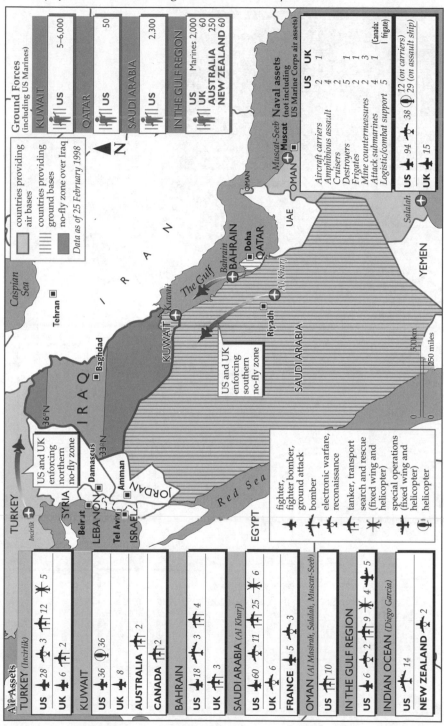

Production and storage sites of Iraqi biological and chemical warfare agents

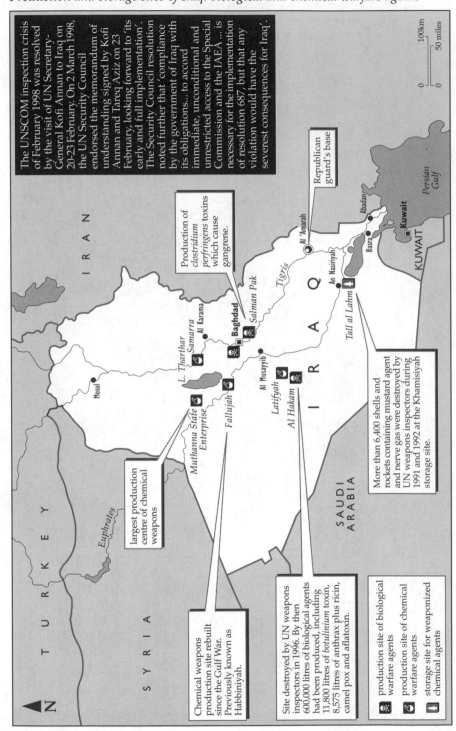

The UNSCOM inspection crisis of February 1998 was resolved by the visit of UN Secretary-General Kofi Annan to Iraq on 20–23 February. On 2 March 1998, the UN Security Council endorsed the memorandum of understanding signed by Kofi Annan and Tareq Aziz on 23 February, looking forward to 'its early and full implementation'. The Security Council resolution noted further that 'compliance by the government of Iraq with its obligations... to accord immediate, unconditional and unrestricted access to the Special Commission and the IAEA ... is necessary for the implementation of resolution 687, but that any violation would have the severest consequences for Iraq'.

Republican guard's base

Production of *clostridium perfringens* toxins which cause gangrene.

More than 6,400 shells and rockets containing mustard agent and nerve gas were destroyed by UN weapons inspectors during 1991 and 1992 at the Khamisiyah storage site.

largest production centre of chemical weapons

Chemical weapons production site rebuilt since the Gulf War. Previously known as Habbiniyah.

Site destroyed by UN weapons inspectors in 1996. By then 600,000 litres of biological agents had been produced, including 11,800 litres of *botulinium* toxin, 8,575 litres of anthrax plus ricin, camel pox and aflatoxin.

IRAN

TURKEY

SYRIA

IRAQ

SAUDI ARABIA

KUWAIT

Persian Gulf

Mosul

L. Tharthar

Samarra
Al Karama

Baghdad

Salman Pak

Tigris

Muthanna State Enterprise

Fallujah

Al Musayyib

Latifiyah

Al Hakam

Al 'Amarah

An Nasiriyah

Tall al Lahm

Abadan

Basra

Kuwait

Euphrates

- production site of biological warfare agents
- production site of chemical warfare agents
- storage site for weaponized chemical agents

100km

50 miles

N

Iraq: toxins, biological and chemical agents (table 1)

Agents	Production method	Effects	Method of delivery	Chemical delivery systems developed by Iraq
Mustard	A blister agent easily synthesised from commonly available chemicals.	Causes blisters that then burst, often become infected and may take weeks to heal. Also causes chronic respiratory complaints.	Airborne dispersal: aerosol. Alternatively, thickened in viscous form for persistent effect.	Aerial bombs, rockets and artillery shells
HCN	A blood agent synthetically produced hydrogen cyanide.	Blocks the uptake of oxygen from the blood. Main symptoms are breathing difficulties, convulsions and death due to respiratory failure.	Airborne: gaseous release and diffusion	Unproven (suspected use)
Phosgene	A choking agent: synthetically produced carbonyl dichloride.		Airborne: gaseous release and diffusion	Unproven
Sarin	A nerve agent: an organophosphorus chemical compound similar in chemistry to an insecticide. Produced synthetically	Acts by blocking normal nerve function. Kills rapidly through paralysis of the respiratory muscles (often complicated effects on the cardiac, nervous and gastro-intestinal systems).	As mustard but with lower vapour pressure and hence disperses more quickly.	Missile warheads, rockets, aerial bombs and field artillery
Tabun	A nerve agent: as sarin	As sarin	As sarin	As sarin
VX	A nerve agent: as sarin, but more potent and more persistent.	As sarin	Delivered in thickened or dry form for persistent effects.	
BZ	A psychotropic agent: quinuclidinyl benzilate is a synthetically produced hallucinogen.	Mind altering drug: incapacitating agent	Airborne: gaseous release and diffusion	R&D

Iraq is known to have used chemical weapons on the following occasions: 1983, Iran–Iraq war (mustard); 1987, Iran–Iraq war (sarin, tabun and mustard used on Al Faw Peninsula); 1987, attack on Kurdish village of Halbaja (agents variously referred to as phosgene, hydrogen cyanide, mustard, sarin and tabun).

UNSCOM has destroyed or rendered harmless more than 480,000 litres of chemical warfare agents, including mustard agent and nerve agents, sarin and tabun; more than 28,000 filled and 12,000 empty chemical weapons; roughly 1.8m litres, more than 1m kg and 648 barrels of 45 different precursor chemicals for the production of chemical weapons.

Iraq: toxins, biological and chemical agents (table 2)

Agents	Production method	Effects	Method of delivery	BW-filled and deployed delivery systems claimed destroyed by Iraq (13 February 1998)	
Anthrax (biological)	Spores from *bacillus anthracis* bacterium	Severe pneumonia-like illness, with death in 1–5 days through poisoning of organs and blood stream	Airborne dispersal: particulate aerosol	Missile warheads Aerial bombs Aircraft aerosol spray	5 50 4
Botulinium toxin	Toxin derived from *clostridium botulinum*	Symptoms include dry mouth, difficulty with vision, speech and swallowing, nausea, vomiting and dizziness. Death occurs in hours from progressive muscular paralysis and respiratory failure.	Airborne dispersal: particulate aerosol	Missile warheads Aerial bombs	16 100
Gas gangrene	Derived from *clostridium perfringens*	Designed to cause gas gangrene in combination with shrapnel weapons through wounds.		R&D	
Aflatoxin	Derived from *aspergillus flavus* and *aspergillus parasticus*	Attacks the immune system and is carcinogenic. Short-term incapacitant	Particulate aerosol	Missile warheads Aerial bombs Field artillery	4 7
Ricin	Toxin derived from castor bean plant	Causes a severe breakdown of lung tissue resulting in haemorrhagic pneumonia and death.	Airborne dispersal: particulate aerosol	R&D	

The Oslo 2 Agreement

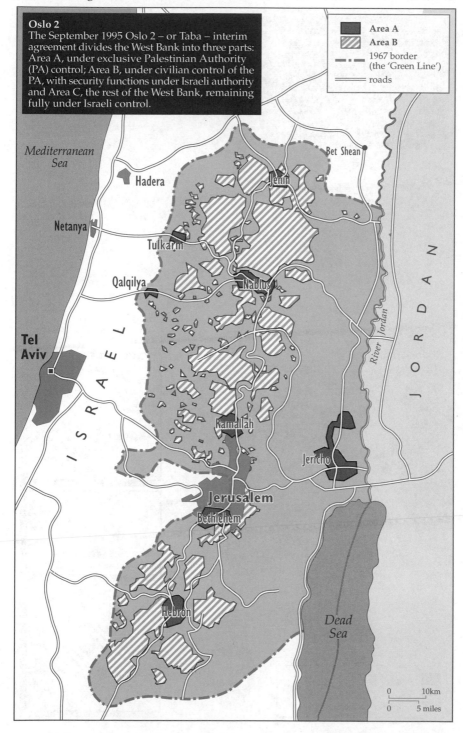

Oslo 2
The September 1995 Oslo 2 – or Taba – interim agreement divides the West Bank into three parts: Area A, under exclusive Palestinian Authority (PA) control; Area B, under civilian control of the PA, with security functions under Israeli authority and Area C, the rest of the West Bank, remaining fully under Israeli control.

Area A
Area B
1967 border (the 'Green Line')
roads

Mediterranean Sea

Bet Shean

Hadera

Jenin

Netanya

Tulkarm

Qalqilya

Nablus

Tel Aviv

ISRAEL

JORDAN

River Jordan

Ramallah

Jericho

Jerusalem

Bethlehem

Hebron

Dead Sea

0   10km
0   5 miles

Israeli redeployments on the West Bank

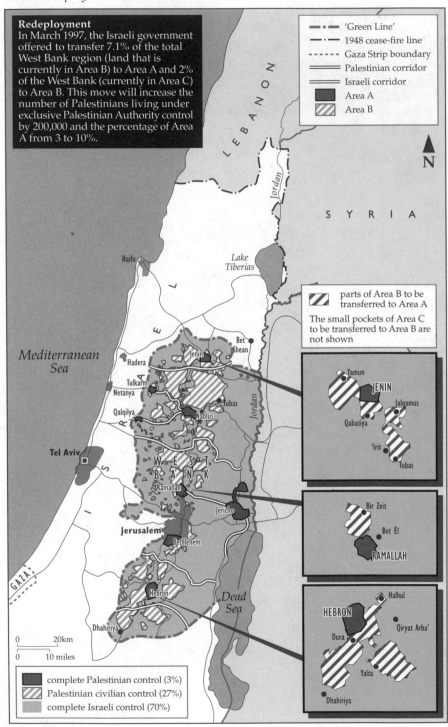

Redeployment
In March 1997, the Israeli government offered to transfer 7.1% of the total West Bank region (land that is currently in Area B) to Area A and 2% of the West Bank (currently in Area C) to Area B. This move will increase the number of Palestinians living under exclusive Palestinian Authority control by 200,000 and the percentage of Area A from 3 to 10%.

'Green Line'
1948 cease-fire line
Gaza Strip boundary
Palestinian corridor
Israeli corridor
Area A
Area B

parts of Area B to be transferred to Area A

The small pockets of Area C to be transferred to Area B are not shown

LEBANON

SYRIA

Jordan

Lake Tiberias

Haifa

Bet Shean

Mediterranean Sea

Hadera

Jenin

Tulkarm
Netanya

Tubas

Qalqilya

Nablus

Tel Aviv

I S R A E L

W E S T   B A N K

Jordan

Ramallah

Jericho

Jerusalem

Bethlehem

Hebron

Dead Sea

Dhahiriya

GAZA

JENIN
Yamun
Jalqamus
Qabatiya
'Irit
Tubas

RAMALLAH
Bir Zeit
Bet Él

HEBRON
Halhul
Qiryat Arba'
Dura
Yalta
Dhahiriya

0	20km
0	10 miles

complete Palestinian control (3%)
Palestinian civilian control (27%)
complete Israeli control (70%)

Jerusalem: existing and planned settlements and built-up areas

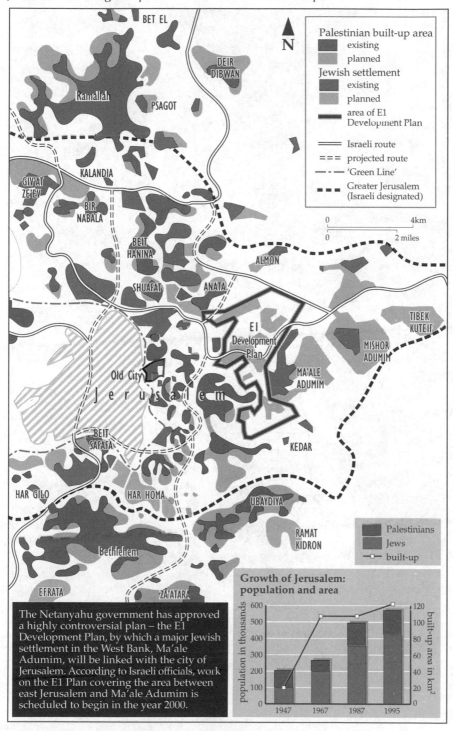

N

Palestinian built-up area
- existing
- planned

Jewish settlement
- existing
- planned

area of E1 Development Plan

Israeli route

projected route

'Green Line'

Greater Jerusalem (Israeli designated)

| 0 | 4km |
| 0 | 2 miles |

BET EL

DEIR DIBWAN

Ramallah

PSAGOT

GIV'AT ZE'EV

KALANDIA

BIR NABALA

BEIT HANINA

ALMON

SHUAFAT

ANATA

E1 Development Plan

TIBEK KUTEIF

MISHOR ADUMIM

Old City

Jerusalem

MA'ALE ADUMIM

BEIT SAFAFA

KEDAR

HAR GILO

HAR HOMA

UBAYDIYA

RAMAT KIDRON

Bethlehem

EFRATA

ZA'ATARA

The Netanyahu government has approved a highly controversial plan – the E1 Development Plan, by which a major Jewish settlement in the West Bank, Ma'ale Adumim, will be linked with the city of Jerusalem. According to Israeli officials, work on the E1 Plan covering the area between east Jerusalem and Ma'ale Adumim is scheduled to begin in the year 2000.

Palestinians
Jews
built-up

Growth of Jerusalem: population and area

population in thousands

built-up area in km²

1947 1967 1987 1995

Israel and water resources on the West Bank

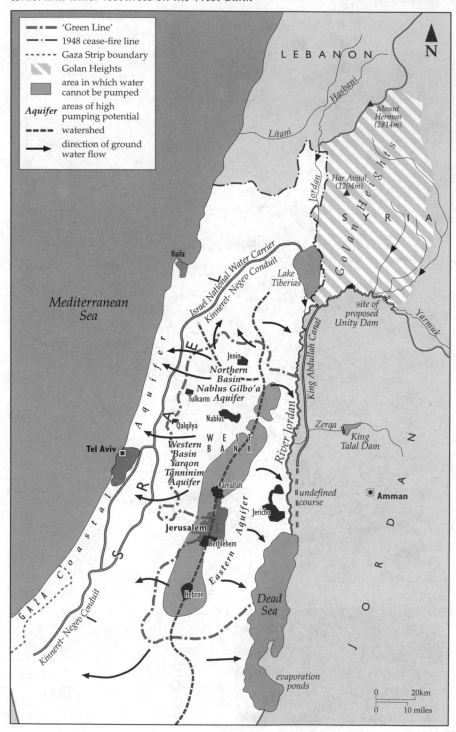

IDF/SLA and Hizbollah clashes in south Lebanon, 1997

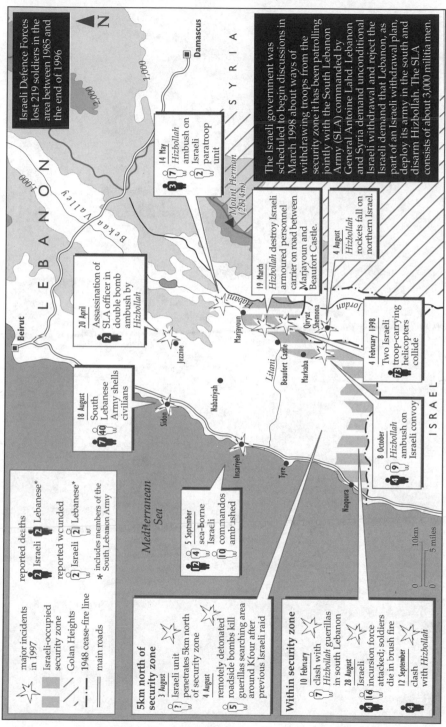

Oil production and reserves world-wide

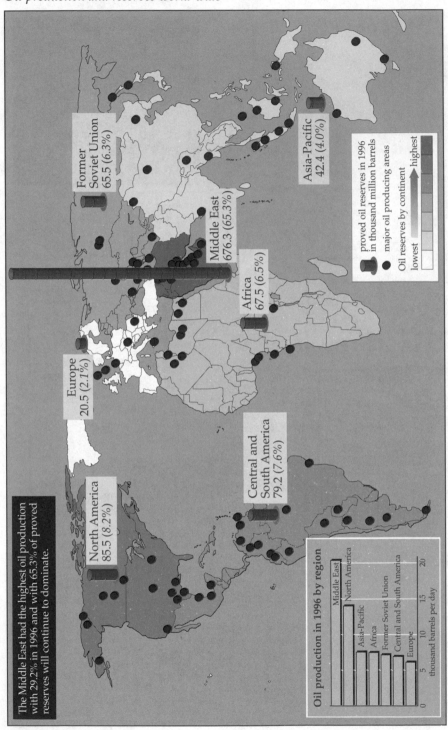

The Middle East had the highest oil production with 29.2% in 1996 and with 65.3% of proved reserves will continue to dominate.

Former Soviet Union 65.5 (6.3%)

Asia-Pacific 42.4 (4.0%)

Middle East 676.3 (65.3%)

Africa 67.5 (6.5%)

Europe 20.5 (2.1%)

Central and South America 79.2 (7.6%)

North America 85.5 (8.2%)

proved oil reserves in 1996 in thousand million barrels

major oil producing areas

Oil reserves by continent

lowest — highest

Oil production in 1996 by region

Middle East
North America
Asia-Pacific
Africa
Former Soviet Union
Central and South America
Europe

thousand barrels per day

0   5   10   15   20

Natural gas production and reserves world-wide

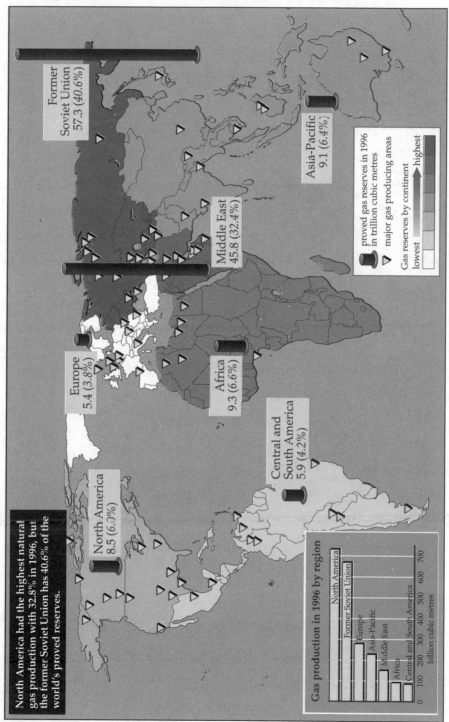

Former Soviet Union 57.3 (40.6%)

Asia-Pacific 9.1 (6.4%)

Middle East 45.8 (32.4%)

Europe 5.4 (3.8%)

Africa 9.3 (6.6%)

Central and South America 5.9 (4.2%)

North America 8.5 (6.0%)

North America had the highest natural gas production with 32.8% in 1996, but the former Soviet Union has 40.6% of the world's proved reserves.

proved gas reserves in 1996 in trillion cubic metres

major gas producing areas

Gas reserves by continent

lowest   highest

Gas production in 1996 by region

North America
Former Soviet Union
Europe
Asia-Pacific
Middle East
Africa
Central and South America

0   100 200 300 400 500 600 700
billion cubic metres

Proposed and existing pipelines from the Caspian Basin

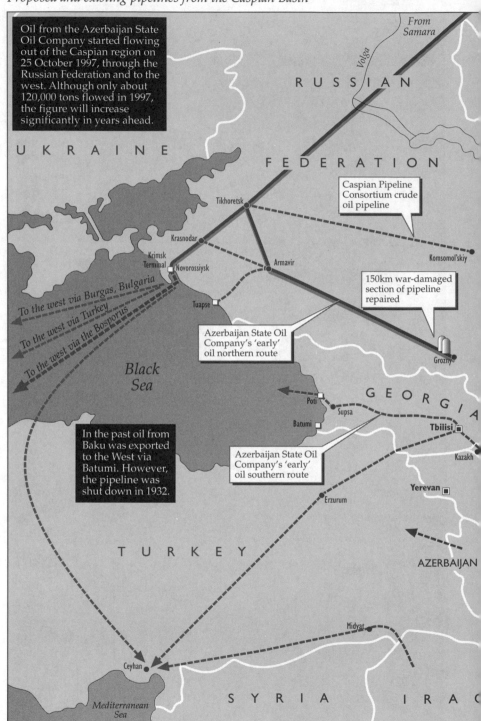

Oil from the Azerbaijan State Oil Company started flowing out of the Caspian region on 25 October 1997, through the Russian Federation and to the west. Although only about 120,000 tons flowed in 1997, the figure will increase significantly in years ahead.

Caspian Pipeline Consortium crude oil pipeline

150km war-damaged section of pipeline repaired

Azerbaijan State Oil Company's 'early' oil northern route

In the past oil from Baku was exported to the West via Batumi. However, the pipeline was shut down in 1932.

Azerbaijan State Oil Company's 'early' oil southern route

From Samara

Volga

R U S S I A N

U K R A I N E

F E D E R A T I O N

Tikhoretsk

Krasnodar

Komsomol'skiy

Krimsk Terminal

Novorossiysk

Armavir

Tuapse

Grozny

To the west via Burgas, Bulgaria

To the west via Turkey

To the west via the Bosporus

Black Sea

Poti

Supsa

G E O R G I A

Batumi

Tbilisi

Kazakh

Yerevan

Erzurum

AZERBAIJAN

T U R K E Y

Midyat

Ceyhan

Mediterranean Sea

S Y R I A

I R A Q

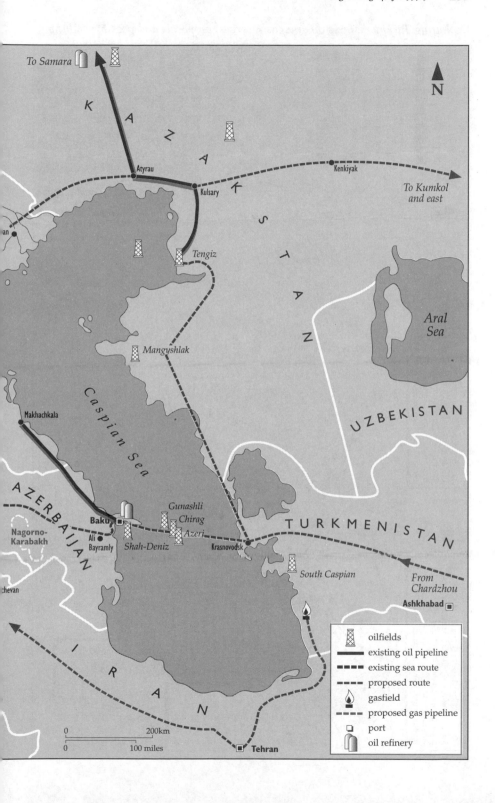

Exploiting Turkmenistan's oil and gas reserves: proposals and possible solutions

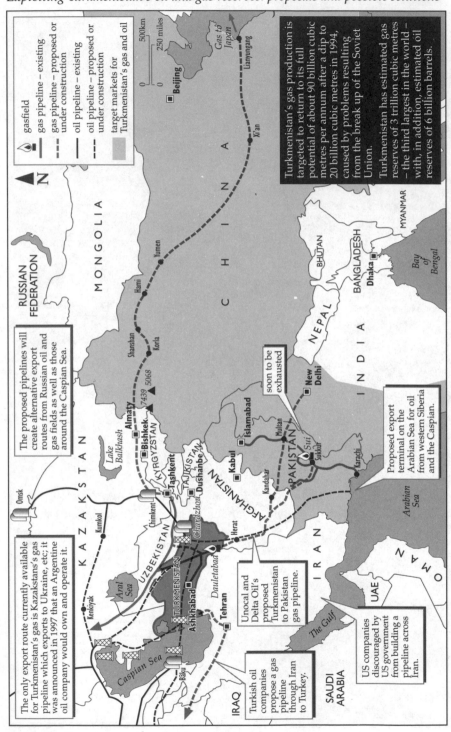

Legend:

- gasfield
- gas pipeline – existing
- gas pipeline – proposed or under construction
- oil pipeline – existing
- oil pipeline – proposed or under construction
- target markets for Turkmenistan's gas and oil

Turkmenistan's gas production is targeted to return to its full potential of about 90 billion cubic metres per annum after a dip to 20 billion cubic metres in 1994, caused by problems resulting from the break up of the Soviet Union.

Turkmenistan has estimated gas reserves of 3 trillion cubic metres – the third largest in the world – with, in addition, estimated oil reserves of 6 billion barrels.

The proposed pipelines will create alternative export routes from Russian oil and gas fields as well as those around the Caspian Sea.

The only export route currently available for Turkmenistan's gas is Kazakstan's gas pipeline which exports to Ukraine, etc; it was announced in 1997 that an Argentine oil company would own and operate it.

soon to be exhausted

Proposed export terminal on the Arabian Sea for oil from western Siberia and the Caspian.

Unocal and Delta Oil's proposed Turkmenistan to Pakistan gas pipeline.

Turkish oil companies propose a gas pipeline through Iran to Turkey.

US companies discouraged by US government from building a pipeline across Iran.

Gas to Japan

Oil and gas pipelines through Ukraine

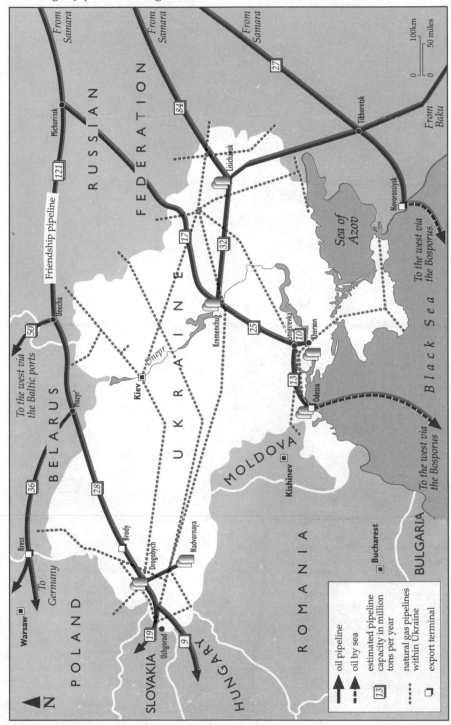

French force reductions in Africa

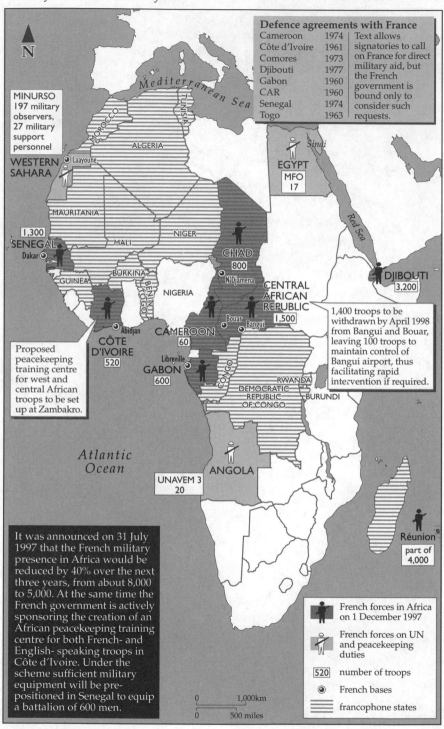

N

Defence agreements with France

Cameroon	1974	Text allows
Côte d'Ivoire	1961	signatories to call
Comores	1973	on France for direct
Djibouti	1977	military aid, but
Gabon	1960	the French
CAR	1960	government is
Senegal	1974	bound only to
Togo	1963	consider such
		requests.

MINURSO
197 military
observers,
27 military
support
personnel

WESTERN
SAHARA

Laayoune

Mediterranean Sea

MOROCCO

TUNISIA

ALGERIA

EGYPT
MFO
17

Sinai

Red Sea

MAURITANIA

NIGER

1,300
SENEGAL
Dakar

MALI

CHAD
800
N'Djamena

DJIBOUTI
3,200

GUINEA

BURKINA

BENIN TOGO

NIGERIA

CENTRAL
AFRICAN
REPUBLIC
1,500

Bouar   Bangui

1,400 troops to be
withdrawn by April 1998
from Bangui and Bouar,
leaving 100 troops to
maintain control of
Bangui airport, thus
facilitating rapid
intervention if required.

Abidjan   CAMEROON
60

CÔTE
D'IVOIRE
520

Libreville

GABON
600

CONGO

RWANDA
DEMOCRATIC
REPUBLIC
OF CONGO   BURUNDI

Proposed
peacekeeping
training centre
for west and
central African
troops to be set
up at Zambakro.

*Atlantic
Ocean*

ANGOLA

UNAVEM 3
20

Réunion
part of
4,000

It was announced on 31 July
1997 that the French military
presence in Africa would be
reduced by 40% over the next
three years, from about 8,000
to 5,000. At the same time the
French government is actively
sponsoring the creation of an
African peacekeeping training
centre for both French- and
English- speaking troops in
Côte d'Ivoire. Under the
scheme sufficient military
equipment will be pre-
positioned in Senegal to equip
a battalion of 600 men.

French forces in Africa
on 1 December 1997

French forces on UN
and peacekeeping
duties

520   number of troops

French bases

francophone states

0   1,000km

0   500 miles

Changes announced to US military presence on Okinawa

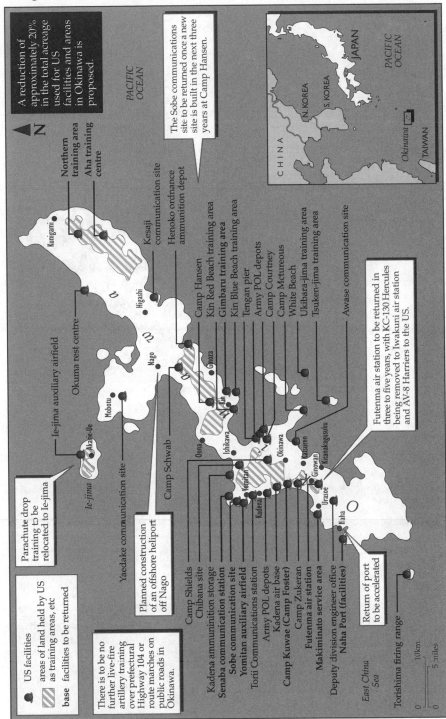

The Kurile Islands dispute

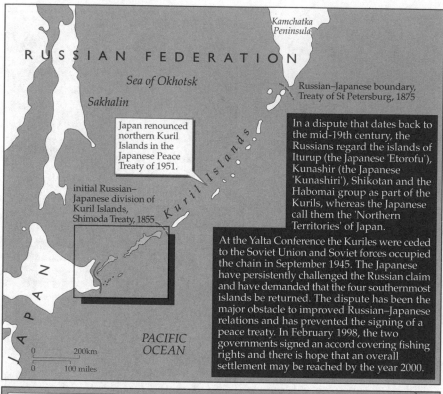

RUSSIAN FEDERATION

Kamchatka Peninsula

Sea of Okhotsk

Sakhalin

Russian–Japanese boundary, Treaty of St Petersburg, 1875

Japan renounced northern Kuril Islands in the Japanese Peace Treaty of 1951.

Kuril Islands

initial Russian–Japanese division of Kuril Islands, Shimoda Treaty, 1855

J A P A N

PACIFIC OCEAN

0   200km
0   100 miles

In a dispute that dates back to the mid-19th century, the Russians regard the islands of Iturup (the Japanese 'Etorofu'), Kunashir (the Japanese 'Kunashiri'), Shikotan and the Habomai group as part of the Kurils, whereas the Japanese call them the 'Northern Territories' of Japan.

At the Yalta Conference the Kuriles were ceded to the Soviet Union and Soviet forces occupied the chain in September 1945. The Japanese have persistently challenged the Russian claim and have demanded that the four southernmost islands be returned. The dispute has been the major obstacle to improved Russian–Japanese relations and has prevented the signing of a peace treaty. In February 1998, the two governments signed an accord covering fishing rights and there is hope that an overall settlement may be reached by the year 2000.

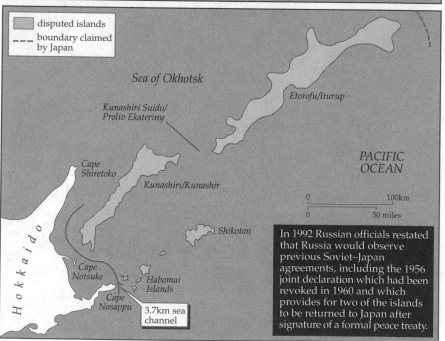

▢ disputed islands
--- boundary claimed by Japan

Sea of Okhotsk

Etorofu/Iturup

Kunashiri Suido/ Proliv Ekateriny

Cape Shiretoko

Kunashiri/Kunashir

PACIFIC OCEAN

0   100km
0   50 miles

Shikotan

Cape Notsuke

Habomai Islands

Cape Nosappu

3.7km sea channel

H o k k a i d o

In 1992 Russian officials restated that Russia would observe previous Soviet–Japan agreements, including the 1956 joint declaration which had been revoked in 1960 and which provides for two of the islands to be returned to Japan after signature of a formal peace treaty.

The Russian–Chinese border

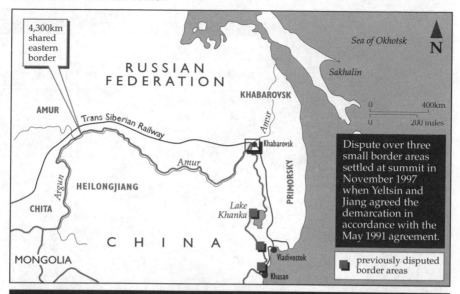

4,300km shared eastern border

RUSSIAN FEDERATION

AMUR

Trans Siberian Railway

HEILONGJIANG

CHITA

MONGOLIA

CHINA

KHABAROVSK

Sea of Okhotsk

Sakhalin

N

Amur

Khabarovsk

Amur

Argun

PRIMORSKY

Lake Khanka

Vladivostok

Khasan

0   400km
0   200 miles

Dispute over three small border areas settled at summit in November 1997 when Yeltsin and Jiang agreed the demarcation in accordance with the May 1991 agreement.

■ previously disputed border areas

At the fifth Sino-Soviet summit held in Beijing on 9–11 November 1997, agreement was reached on a declaration that finally settled disagreements relating to the implementation of the accord reached in 1991 that mapped out the entire border area between the two countries. It was also agreed that a separate agreement, to be negotiated, will cover the joint use of islands along the border rivers of Amur and Ussuri. The agreement is accompanied by the introduction of military confidence-building measures in the border area, including the reduction of troop levels.

Boundary following line formed by the main current in the river and not following the Chinese river bank.

KHABAROVSK

Trans-Siberian Railway

Ostrov Tarabarovskiy

Boundary claimed by Chinese.

Amur River

Khabarovsk

Fuyan

Ostrov Bol'shoy Ussuriyskiy

Kazakevicheva Channel

Boundary claimed by Russians.

Ussuri

Kazakevichevo

Disputed area 175km²

HEILONGJIANG

RUSSIAN FEDERATION

CHINA

0   10km
0   5 miles

Refugees and internally displaced people

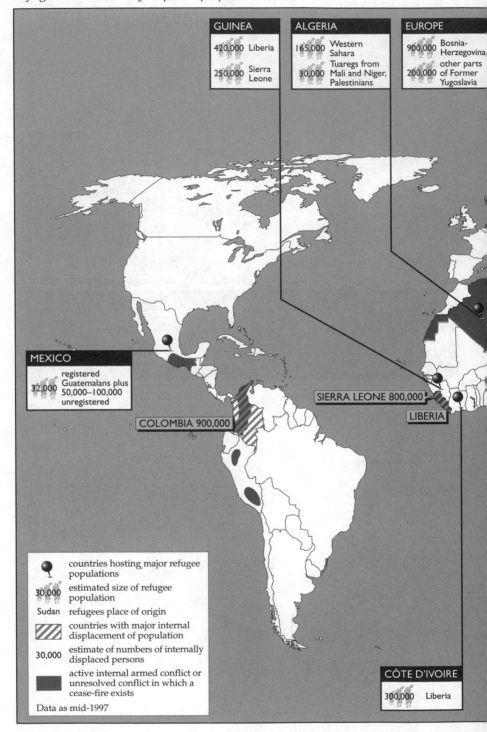

GUINEA

420,000 Liberia

250,000 Sierra Leone

ALGERIA

165,000 Western Sahara

30,000 Tuaregs from Mali and Niger, Palestinians

EUROPE

900,000 Bosnia-Herzegovina

200,000 other parts of Former Yugoslavia

MEXICO

32,000 registered Guatemalans plus 50,000–100,000 unregistered

COLOMBIA 900,000

SIERRA LEONE 800,000

LIBERIA

countries hosting major refugee populations

30,000 estimated size of refugee population

Sudan refugees place of origin

countries with major internal displacement of population

30,000 estimate of numbers of internally displaced persons

active internal armed conflict or unresolved conflict in which a cease-fire exists

Data as mid-1997

CÔTE D'IVOIRE

300,000 Liberia

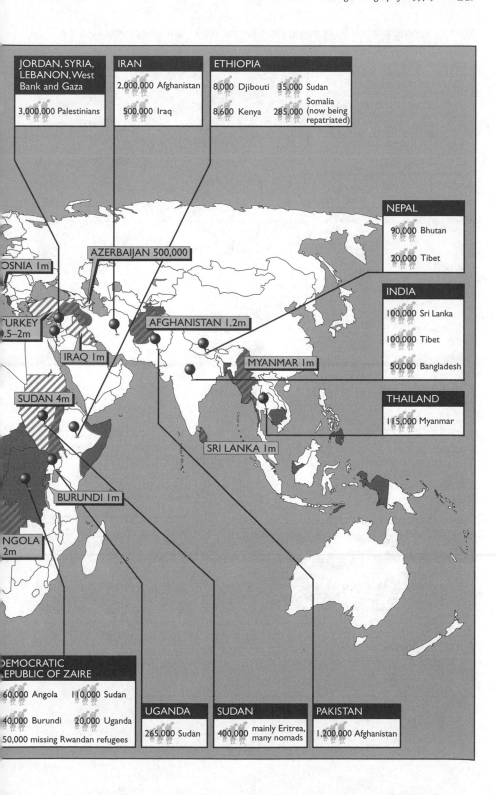

JORDAN, SYRIA, LEBANON, West Bank and Gaza

3,000,000 Palestinians

IRAN

2,000,000 Afghanistan

500,000 Iraq

ETHIOPIA

8,000 Djibouti   35,000 Sudan

8,600 Kenya   285,000 Somalia (now being repatriated)

NEPAL

90,000 Bhutan

20,000 Tibet

INDIA

100,000 Sri Lanka

100,000 Tibet

50,000 Bangladesh

THAILAND

115,000 Myanmar

BOSNIA 1m

AZERBAIJAN 500,000

TURKEY 0.5–2m

IRAQ 1m

AFGHANISTAN 1.2m

MYANMAR 1m

SUDAN 4m

SRI LANKA 1m

BURUNDI 1m

ANGOLA 2m

DEMOCRATIC REPUBLIC OF ZAIRE

60,000 Angola   110,000 Sudan

40,000 Burundi   20,000 Uganda

50,000 missing Rwandan refugees

UGANDA

265,000 Sudan

SUDAN

400,000 mainly Eritrea, many nomads

PAKISTAN

1,200,000 Afghanistan

Refugee populations and movements in Central Africa and the Horn of Africa

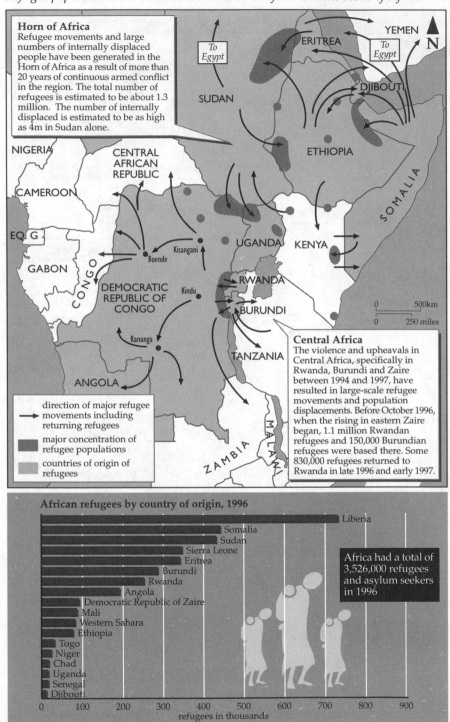

Horn of Africa
Refugee movements and large numbers of internally displaced people have been generated in the Horn of Africa as a result of more than 20 years of continuous armed conflict in the region. The total number of refugees is estimated to be about 1.3 million. The number of internally displaced is estimated to be as high as 4m in Sudan alone.

Central Africa
The violence and upheavals in Central Africa, specifically in Rwanda, Burundi and Zaire between 1994 and 1997, have resulted in large-scale refugee movements and population displacements. Before October 1996, when the rising in eastern Zaire began, 1.1 million Rwandan refugees and 150,000 Burundian refugees were based there. Some 830,000 refugees returned to Rwanda in late 1996 and early 1997.

→ direction of major refugee movements including returning refugees

■ major concentration of refugee populations

countries of origin of refugees

African refugees by country of origin, 1996

Liberia
Somalia
Sudan
Sierra Leone
Eritrea
Burundi
Rwanda
Angola
Democratic Republic of Zaire
Mali
Western Sahara
Ethiopia
Togo
Niger
Chad
Uganda
Senegal
Djibouti

Africa had a total of 3,526,000 refugees and asylum seekers in 1996

0   100   200   300   400   500   600   700   800   900
refugees in thousands

Source countries and trafficking routes: the Americas

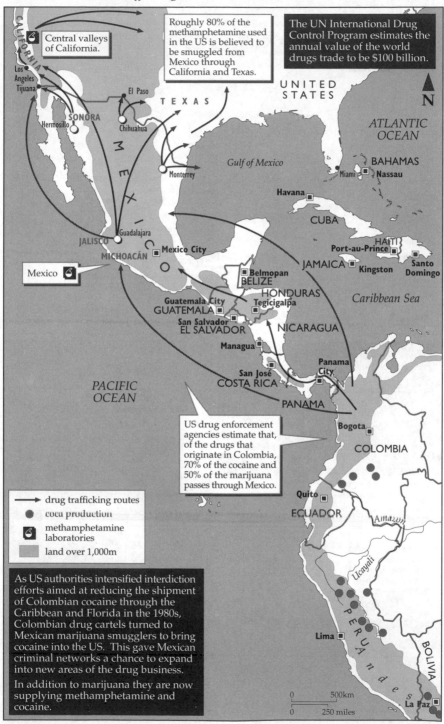

Central valleys of California.

Roughly 80% of the methamphetamine used in the US is believed to be smuggled from Mexico through California and Texas.

The UN International Drug Control Program estimates the annual value of the world drugs trade to be $100 billion.

Mexico

US drug enforcement agencies estimate that, of the drugs that originate in Colombia, 70% of the cocaine and 50% of the marijuana passes through Mexico.

CALIFORNIA
Los Angeles
Tijuana
El Paso
TEXAS
SONORA
Hermosillo
Chihuahua
Monterrey
Guadalajara
JALISCO
MICHOACÁN
Mexico City
UNITED STATES
N
ATLANTIC OCEAN
Gulf of Mexico
BAHAMAS
Miami
Nassau
Havana
CUBA
Port-au-Prince
HAITI
JAMAICA
Kingston
Santo Domingo
M E X I C O
Belmopan
BELIZE
HONDURAS
Tegicigalpa
Caribbean Sea
Guatemala City
GUATEMALA
San Salvador
EL SALVADOR
Managua
NICARAGUA
San José
COSTA RICA
Panama City
PANAMA
PACIFIC OCEAN
Bogota
COLOMBIA
Quito
ECUADOR
Amazon
Lima
Ucayali
P E R U
A n d e s
BOLIVIA
La Paz

drug trafficking routes
coca production
methamphetamine laboratories
land over 1,000m

As US authorities intensified interdiction efforts aimed at reducing the shipment of Colombian cocaine through the Caribbean and Florida in the 1980s, Colombian drug cartels turned to Mexican marijuana smugglers to bring cocaine into the US. This gave Mexican criminal networks a chance to expand into new areas of the drug business.

In addition to marijuana they are now supplying methamphetamine and cocaine.

0   500km
0   250 miles

Source countries and trafficking routes: Central Asia and South-east Asia

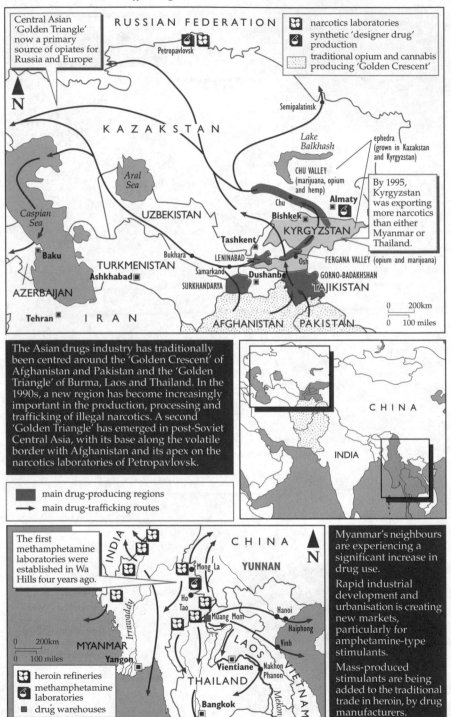

Economic groupings in the Americas

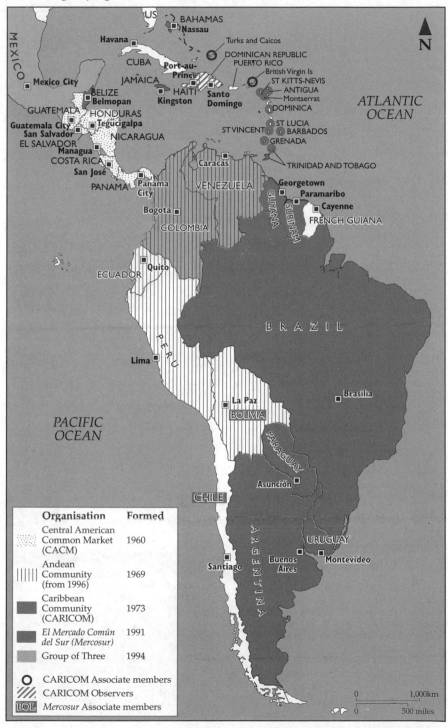

Organisation	Formed
Central American Common Market (CACM)	1960
Andean Community (from 1996)	1969
Caribbean Community (CARICOM)	1973
El Mercado Común del Sur (Mercosur)	1991
Group of Three	1994

○ CARICOM Associate members

▨ CARICOM Observers

BOL Mercosur Associate members

Algeria: the zone of death

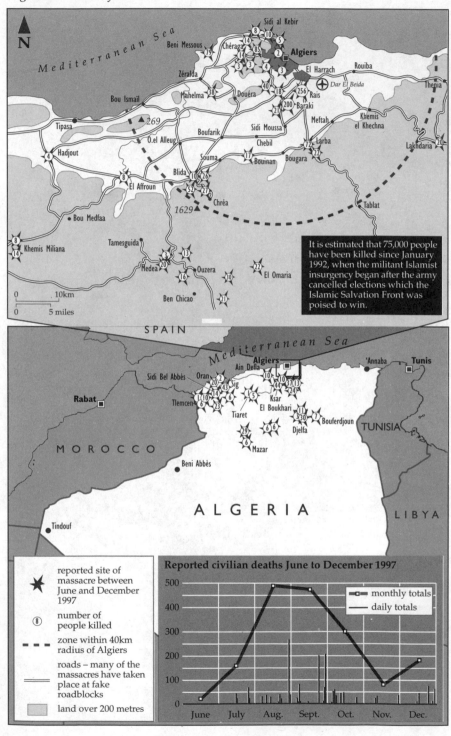

It is estimated that 75,000 people have been killed since January 1992, when the militant Islamist insurgency began after the army cancelled elections which the Islamic Salvation Front was poised to win.

Reported civilian deaths June to December 1997

★ reported site of massacre between June and December 1997

⑧ number of people killed

- - - zone within 40km radius of Algiers

roads – many of the massacres have taken place at fake roadblocks

land over 200 metres